T0280949

Advanced Controls for
Intelligent Buildings

Advanced Controls for Intelligent Buildings

A Holistic Approach for Successful Businesses

Siddharth Goyal

CRC Press
Taylor & Francis Group
Boca Raton London New York

CRC Press is an imprint of the
Taylor & Francis Group, an **informa** business

First edition published 2021
by CRC Press
6000 Broken Sound Parkway NW, Suite 300, Boca Raton, FL 33487-2742

and by CRC Press
2 Park Square, Milton Park, Abingdon, Oxon, OX14 4RN

CRC Press is an imprint of Taylor & Francis Group, LLC

Library of Congress Cataloging-in-Publication Data

Names: Goyal, Siddharth, author.
Title: Advanced controls for intelligent buildings : a holistic approach
for successful businesses / Siddharth Goyal, Ph.D.
Description: Boca Raton : CRC Press, 2021. | Includes bibliographical
references and index.
Identifiers: LCCN 2021001768 (print) | LCCN 2021001769 (ebook) | ISBN
9781032009650 (hardback) | ISBN 9781032009674 (paperback) | ISBN
9781003176589 (ebook)
Subjects: LCSH: Intelligent buildings. | Buildings--Mechanical
equipment--Automatic control. | Building materials industry.
Classification: LCC TH6012 .G69 2021 (print) | LCC TH6012 (ebook) | DDC
696.0285/4678--dc23
LC record available at https://lccn.loc.gov/2021001768
LC ebook record available at https://lccn.loc.gov/2021001769

ISBN: 978-1-032-00965-0 (hbk)
ISBN: 978-1-032-00967-4 (pbk)
ISBN: 978-1-003-17658-9 (ebk)

Typeset in Krantz
by KnowledgeWorks Global Ltd.

*To my wife
and my children.*

Contents

Preface

Controls play an important role in today's technology world with several strings attached to the ancient world. We realize the importance of control and have been making decisions every day that use a form of control although we may not be consciously aware of them. For example, a person deciding the best departing time to go to the office based on certain criteria and constraints (e.g., arrival time before 8:00 am), and updating the route or driving style (conservative or aggressive) according to real-time traffic conditions. From control's perspective, this scenario could be directly related to MPC (Model Predictive Control), in which the user is making decisions based on future predictions and updating the decisions based on the user's deviations from original predictions. Sensing and modeling are also the crucial parts of an overall control system. In the above example, the eyes are acting as sensors to monitor the traffic and other environmental conditions, and a model is created in the brain to estimate the arrival time considering past experiences, i.e., historical data. Similarly, sensing, modeling, and controls play a key role in buildings—the places where we work, live, and spend most of our time.

Buildings have significant impact on not only an individual's life (health, productivity, etc.) but also on the environment and economy. It is critical that we design and develop such systems and applications in buildings intelligently. Therefore, this book focuses on the design, development, and deployment of advanced control systems and applications for building systems. For successful delivery of controls projects and increasing the adoption rate of such products in the market, or increase the acceptance rate of such ideas within an organization, it is important to connect the control applications and use cases to the business needs, market challenges, and the corresponding financial analysis. Therefore, the second part of the book is focused on the business aspects understanding the key stakeholders, market barriers, and challenges to increase the success rate and deliver quality results concurrently. In a nutshell, with a blend of both technical and business perspectives, the book provides a holistic, end-to-end approach toward intelligent building systems to help businesses grow and solve their critical problems.

Organization of the Book

The book is divided into two major parts. The first half of the book targets the technical aspects from design to deployment of control applications and use cases while the second half focuses on business needs and requirements.

Chapter 1 provides an overview of building, the equipment used in buildings and their typical control systems. Chapter 2 describes the commonly-used sensors and sensing techniques in building systems. Modeling methods, trade-offs between the methods, and their influence on control methods are explained in Chapter 3. Control approaches and techniques (including both advanced and prevalent techniques in building systems) and choice of appropriate control framework based on functional, system, and business requirements are described in Chapter 4. Once the control applications are designed and developed, they need to be tested and deployed on real systems. Chapter 5 presents deployment mechanisms, including software architecture, for the control applications.

Chapter 6 discusses the technical uses cases from business point of view along with Internet of Things (IoT) and Artificial Intelligence (AI) topics. Chapter 6 provides a smooth transition to segue into second half of the book. Chapter 7 helps readers understand the distribution channels and various stakeholders involved during the entire value chain, and their influence during multiple product stages. Business models, industry structure, and marketing aspects of a business, including the non-technical challenges that need to be overcome before launching a product, are explained in Chapter 8. Chapter 9 discusses the financial factors, return on investment, and the risks that need to be evaluated while deciding the strategy of a new product. Finally, Chapter 10 summarizes and concludes the book with an overall framework and recommendations for the readers.

Audience and Benefits to the Readers

The primary audience of the book are engineers, researchers, and early-career business leaders, who are interested in understanding the technical and business aspects involved in the building controls industry. The intent of the book is to overview multiple topics—instead of drilling down into one specific topic—connecting the dots to provide a holistic perspective. There are several books on individual topics such as controls, modeling. However, there does not exist any book that combines multiple technical concepts (sensing, modeling, estimation, and control) in building systems to provide a strong overview of an end-to-end system. Furthermore, no book connects the technical and the business sides involved during multiple stages in the industry. It is very important for any organization to relate the technical and business aspects, or at least be aware of them, to deliver successful projects and increase the market/industry adoption rate both internally and externally. The book brings the author's work experience in academia, a start-up, a national laboratory, and industry in modeling, estimation, and control of energy and building systems.

The book will help engineers and researchers, who are working on day-to-day activities, and are interested in learning about the bigger problem being solved, and how does their work fit into the problem. In this way, the engineering solutions that are being developed considering business and market challenges are most likely to (1) go past the prototyping stage and being released to the market, (2) possess higher quality and need less re-engineering after the release, and (3) possess higher market and industry adoption rate. The book will also benefit the experienced professionals in research organizations and academia, and new market players (individuals or organizations) in the industry. Early-business leaders will be able to identify innovation ideas and the available techniques with their relation to business challenges and opportunities.

Acknowledgment

Writing a book from conception to end is time-consuming but highly rewarding for me. In addition to clear/concise organization of thoughts and plethora of knowledge, writing a book requires focus, dedication, and motivation for a sustained period of time. I am very blessed to receive the consistent and continuous support from my family. Without the love and support from my wife Lesley, my children Aakash and Reya, my parents Kamla and Ramesh, and brother Shikhar, this book would not have been possible.

Education shapes the career and life of a person. I am fortunate to have a formal education from instructors at three different institutions: (1) Punjab Engineering College for starting and solidifying my interest in control(s) during undergraduate studies; (2) University of Florida for providing me strong technical background and developing analytical, problem-solving, and communication skills; and (3) Marquette University for advancing my management and business skills/knowledge while relating them to the corporate world. I would like to thank the instructors and advisors at these institutions.

It is my pleasure to extend my sincere gratitude to my current employer Divisions Maintenance Group and the former employers (LG Electronics, Pacific Northwest National Laboratory, and Johnson Controls) for providing their support, training, and invaluable experience in the industry. I was fortunate to work for the leading organizations and being able to smoothly transition from academia to applied research to industry. Without the experience from industry, national laboratory, and academia—more importantly their trust to involve me in various projects and products interacting with diversified stakeholders—this book would not have been possible.

I gratefully acknowledge the publisher (Taylor & Francis) and its entire team, in reviewing, editing, and publishing the book including a number of activities before/during/after the publishing process. Finally, I also offer my regards and sincere thanks to all of those who supported me in any respect during the completion of the work, and contributed to the knowledge, education, and experience that I had gained over the past couple of decades.

Author

Dr. Siddharth Goyal was born in Hisar, India. He currently lives with his wife and two children in Atlanta Metropolitan area in Georgia, USA. In his free time, he enjoys running, playing sports (tennis, table tennis, badminton), exploring/traveling, and spending time with his family.

He is currently working as a Director, leading and managing the Data Science team at Divisions Maintenance Group (DMG), Newport, KY. His professional experience ranges from research to product development, product management, strategy, technical leadership, and cross-functional team management. In fortune 500 companies, a software startup, a national laboratory, and several universities. Previous experience includes key positions at LG Electronics, Johnson Controls, Pacific Northwest National Laboratory, University of Florida, National University of Singapore, Indian Institute of Science, Reliance Energy, and GE Motors. In his career, he had worked in multiple domains such as energy systems, intelligent buildings, software applications, artificial intelligence, aircraft and satellites, power plants, and robotics. In the past decade, he has been focusing primarily on solving technical and business problems in the smart buildings industry.

In terms of education, Dr. Goyal holds a Ph.D. and an M.Sc. in Mechanical Engineering from the University of Florida, and a B.E. in Electrical Engineering from Punjab Engineering College. He is also currently enrolled in a part-time MBA program at Marquette University. He published numerous peer-reviewed papers in international conferences and journals. He submitted more than half a dozen patents on new technologies in the area of building controls. He is an active member of ASHRAE, IEEE, and BACnet. He delivered talks at various international conferences and events.

With his diversified experience and education background, he brings an interesting and insightful perspective/content that is highly valuable to the readers of this book.

List of Figures

List of Tables

1

Building Systems

1.1 Introduction and Motivation

Our life is heavily impacted by the buildings—the place where we spend most of our day regardless of the activity. As shown in Figure 1.1, buildings are the major consumers of energy worldwide, contributing to almost 40% of the total energy consumption in the United States [13, Table A-2]. They are also the biggest consumers of electricity, approximately 75% in the US [12]. Therefore, any changes to the buildings (construction, operation, maintenance, usage type, etc.) will affect not only the energy consumption but also the power grid system, including the integration of renewable sources into the system.

Buildings are responsible for almost 30% of the greenhouse emissions [15]. They also have a strong influence on our lives as we spend almost 90% of our time inside buildings [14]. Furthermore, almost $430 billion dollars are spent on buildings every year [13, Table A-2]. A snapshot of the key statistics of buildings in the US is shown in Figure 1.2.

Because of the diversified portfolio of buildings and their high usage by people, any improvement and changes in the buildings have strong individual, social, economic, and environmental impacts. The rest of this chapter discusses the types of buildings, equipment, and systems used in buildings. An overview of smart/intelligent buildings—a potential toward next generation of buildings and an important market trend—is provided at the end of this chapter. Several statistics and facts on buildings are also presented in this chapter as they are important in understanding the market and their impact on business decisions, which are discussed in later chapters.

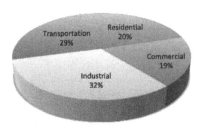 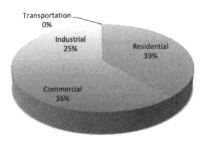

A Energy Consumption in 2017   B Electricity Consumption in 2019

FIGURE 1.1
Annual percentage contribution of total energy and electricity consumption by the sector in the US.

1.2 Building Types

Buildings can be categorized into different types based on activities, usage, demographics, equipment, climate, income, region, occupancy, and size. At the top-level, the buildings are divided into two major types: commercial and residential, which are explained next.

1.2.1 Residential

Residential buildings are primarily used for living purposes. According to the Residential Energy Consumption Survey (RECS) [11], residential buildings are classified into four main categories: single-family detached, single-family attached, apartment (2–4 units and more than 5 units), and mobile homes. The main difference between a single-family detached home and a single-family attached home is that the single-family attached homes share a wall with another home, e.g., townhouse and duplex. On the other hand, a single-family detached home is an independent unit for one household or family with no common wall. The definitions of the home types can be found in RECS [11].

In the US, there are 118.2 million residential buildings with a total of 237.4 billion ft^2 floor area. Percent distribution of the building types based on their quantity and floor area is shown in Figure 1.3. As shown in the figure, single-family detached homes dominate the market in both areas, i.e., quantity and floor area. From a business perspective, it means that a solution (e.g., research or product) targeting single-family detached homes can have much larger and wider impact than a solution designed for other building types in

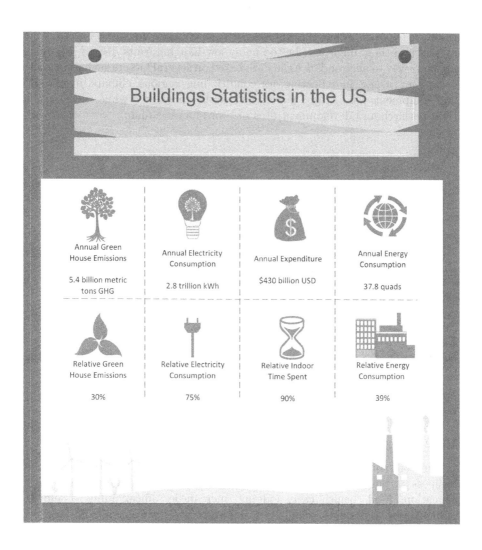

FIGURE 1.2
Key statistics on buildings in the US.

the residential market. This also imposes a higher level of competition and entry-barrier as many individuals and organizations have developed (or have been developing) solutions for that market due to the same reasons. As a result, it is possible that the state-of-art is much more advanced for the same market and there might not be too many low-hanging fruits. RECS [11] offers a unique way to slice and dice the data based on several factors such as income, occupancy, usage, fuel, heating, air-conditioning, regions, climate, insulation, and equipment. All of these factors need to be considered while choosing the right research and development area or a product roadmap.

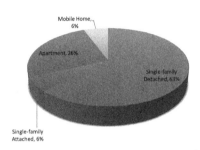
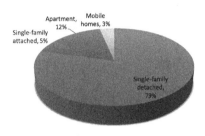

A Number of buildings (118.2 million)   B Floor area of buildings (237.4 billion ft^2)

FIGURE 1.3
Distribution of the quantity and the floor area of residential buildings in the US [11, Table HC10.1].

1.2.2   Commercial

Commercial buildings are the buildings that are used for commerce and business purposes. Commercial buildings are classified into 13 main categories: education, food, health care, lodging, mercantile, office, public assembly, public order and safety, religious worship, service, warehouse and storage, vacant, and other. Definition of the building types and their further classification along with examples can be found in the Commercial Buildings Energy Consumption Survey (CBECS) [10]. For instance, food stores are defined as buildings that are used for retail or wholesale of food. A few examples in this category are grocery stores, food markets, convenience stores, and gas stations with a convenience store. Similar to the residential buildings, relative distribution of the commercial building types based on their quantity and floor area is shown in Figure 1.4.

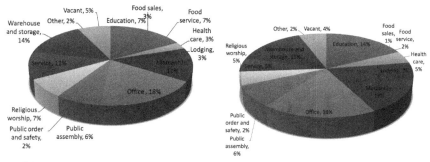

A Number of buildings (5.5 million)   B Floor area of buildings (87.1 billion ft^2)

FIGURE 1.4
Distribution of the quantity (5.5 million [10, Table B3]) and the floor area (87.1 billion ft^2 [10, Table B7]) of commercial buildings in the US.

It is clear from the figure that there is no single building type that dominates the market as opposed to the scenario in residential buildings. Although residential buildings have much higher footprint in terms of floor area, they consume approximately similar energy as compared to commercial buildings. Does it mean that residential buildings are more efficient than commercial buildings or the residences consume less energy because a good amount of people's time is spent in commercial buildings because of work? Moreover, the number of commercial buildings is much less than the residential buildings. Commercial buildings also have higher floor area per building as compared to residential buildings. It implies that one commercial building equates to many residential buildings in some cases. From a business point of view, suppose a new technology or product is launched for commercial buildings that can save 5% energy, which turns out to be 250 kWh monthly savings for one commercial building. To achieve the same level of energy savings, i.e., 250 kWh in a month, the same technology needs to be implemented in 20 homes. Therefore, the sales and marketing strategies for commercial buildings will be different than that of residential buildings. In this case, adoption of the new product in residences requires acceptance of 20 homeowners while adoption of the product in one commercial building requires acceptance only from its building owner but consensus from other stakeholders, which will be discussed in detail in Chapter 7.

A summary of building types and their classification are shown in Figure 1.5. The classification is critical as it helps not only decide the business cases but also determine the products' features, complexity, and testing involved in the entire process. Health-care facilities, as compared to the public assembly buildings, will require higher flexibility in the building systems because of certain constraints and certifications required by the local and federal regulation agencies.

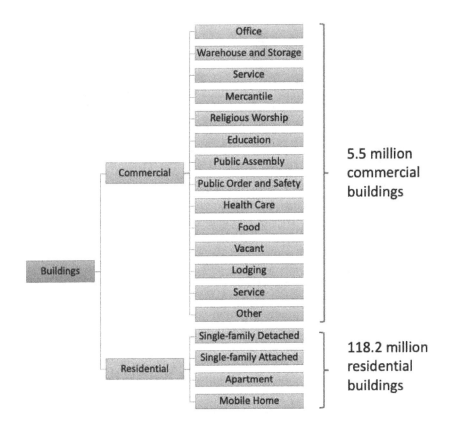

FIGURE 1.5
Building classification tree.

1.3   Building Systems

1.3.1   Overview: System Components and Functionality

Residential and commercial buildings are complex systems with several heterogeneous components that vary quite a bit in terms of functionality, usage, purpose, and architecture (e.g., multiple communication protocols at different communication speeds). A few of these major systems are categorized below:

1. *Heating, Ventilation, and Air-conditioning (HVAC):* As the name suggests, heating and cooling (also referred to as air-conditioning) systems are used to provide heating and cooling, respectively, into a building using the first principles of thermodynamics. Ventilation systems are used to ensure proper indoor air quality (IAQ) inside buildings. Typically, the air quality is managed by using air-filters and controlling the amount of fresh outdoor air. HVAC systems coordinate among each other continuously and act as one system to control the indoor climate. Types and functionality of HVAC systems are explained later in this chapter.

2. *Power:* Power systems are used inside the buildings to generate, supply, and route power between different equipment. Power system components include solar panel, power inverter, battery, electric panel, emergency generators, transformers, meters, fuses, and relays.

3. *Fire Safety:* Fire safety systems detect any fire or smoke related activity in buildings, and take actions accordingly. The actions include notifying the facility managers, operators, and occupants in the buildings via an alarming system. In some advanced cases, multiple mechanisms are used to notify the users, e.g., phone call, text messaging, integration into the building automation system (BAS), and notifications over the Internet. At the same time, the fire suppression system gets activated automatically to extinguish or limit the scope of fire. A few examples of fire suppression systems are water sprinkler systems and gas or chemical based systems.

4. *Security:* These systems are put in place to prevent security breaches and ensure safety/security of assets (people, equipment, properties, data, etc.) in buildings. Access control systems make certain that only authorized people are permitted or denied in specific areas of a building. For example, electrical laboratories require high-voltage and safety training before an employee can use the facility. Surveillance systems capture live video and audio streams from different parts of buildings, mostly remotely, to validate access control and prevent security breaches and intrusions. Video systems can be used to detect employee thefts and to verify actions and events, e.g., check if safety protocols and operating procedures are being followed in emergency situations.

5. *Lighting:* Lighting systems are used to control the lighting in a specific area (either an individual light fixture or a group of lights in a room) in a building including the exterior lights in

parking lots. Controlling a light fixture includes modulating its intensity, changing its color scheme or focus point, or turning it on or off. Dimming and binary control are most commonly used in lighting control systems. A few types of light bulbs are LED (light emitting diodes), fluorescent, incandescent, halogen, and high-intensity discharge [9]. With the use of higher-efficient light bulbs such as LED and compact fluorescent, the energy consumed for lighting has been decreasing at a faster rate than the other building systems. From 2003 to 2012, the lighting energy consumption had decreased by 45%. In 2021, they contribute to 10% buildings energy consumption as compared to 21% in 2003.

6. *In-building Transportation:* Elevators, escalators, and moving walkways fall into this category of systems. These systems move people and items from one place to another inside large buildings, e.g., escalators in malls, elevators in high-rise buildings, and moving walkways in airports. Typically, in-building transportation systems are not directly linked to other building systems, except that there might be a unified display screen for monitoring purposes. However, in some cases depending on the building applications, they could be a part of BAS and/or security systems. Energy consumed by the in-building transportation systems is quite low in comparison with other building systems. However, they can be used for tracking purposes yielding a higher level of user convenience. A business case along those lines is coordination of elevators for large-space events using occupancy monitoring devices.

7. *Electronics and Plug Loads:* Increasingly use of computers, phones, televisions, screens, and other electronics devices had increased their energy consumption from 2% in 2003 to 6% in 2012, and electricity consumption from 4% in 2003 to 10% in 2012 in the commercial sector. Similar trends can be observed for many plug loads such as monitors, printers, desk lamps, projectors, and other miscellaneous devices. Usually, there is no dedicated control system to manage the energy consumption or usage of plug-in loads. From controls perspective, to coordinate and manage the entire building system, the plug loads and electronics are considered as uncertainties in the overall system which have been increasing in magnitude over years. However, entry of wireless power adapters had allowed the users to monitor and control the devices from their smartphones.

8. *Refrigeration:* Refrigeration systems cool a space or substance below the room temperature. Refrigeration systems are also used to control the space temperature by exchanging energy with HVAC systems. Here, refrigeration systems are limited to be contained inside an equipment that maintain below-room temperature for storing and preserving items/goods. Common examples of such

systems are walk-in freezers, refrigerators, grocery freezers, and vending machines. Energy and electricity usage of refrigeration systems in residential buildings are much higher than that of commercial buildings as most residences have at least one refrigerator or freezer.

9. *Others:* There are several other equipment and systems in buildings that are not explained above. A few of them are hot water system for domestic water usage (not HVAC), cooking equipment, washing equipment primarily in residential buildings, parking structure, and IT systems in commercial buildings to manage networking infrastructure, process, software, and data systems. Cybersecurity is usually considered as a part of IT systems.

HVAC systems are the major consumers (almost 40%) of energy in both residential and commercial buildings [13, Table A-2]. As explained in earlier sections, they also have significant impact on people, society, and environment in many ways, the examples and technical topics in this book are focused on HVAC systems. Although HVAC systems are used as a domain application, many technical details such as high-level architecture, sensing, control fundamentals can be applied to the other building systems with some general assumptions. Furthermore, the business concepts and most of the conclusions including the methodologies and approaches are directly applicable to other building systems.

Today, most building systems (lighting, HVAC, power etc.) operate in silos with very limited interactions between each other. Because there are several opportunities and business cases in coordinating such systems, it is important to understand the working of other systems as well. Therefore, in addition to the detailed information on HVAC systems, other systems are also occasionally touched upon at different levels based on the scenario and use case.

1.3.2 HVAC Systems

Working and types of several HVAC systems in buildings are explained in this section. Schematic of an entire HVAC system used in a large commercial building is shown in Figure 1.6. It is an AHU-VAV (air-handling unit and variable-air-volume) system with water-cooled chiller and gas-fired boiler. Its working diagram is simplified for the sake of better understanding.

In this system, the air handling unit delivers conditioned air at 55 °F (12.78 °C) to VAV boxes, which are assigned to each zone. Zone is an area inside a building, which can be a single room or a collection of areas/rooms inside the building. A VAV box is a type of terminal unit, which consists of dampers and a heating mechanism (electric or fluid based). Dampers and heating mechanism are controlled together to deliver the right amount of air at right temperature

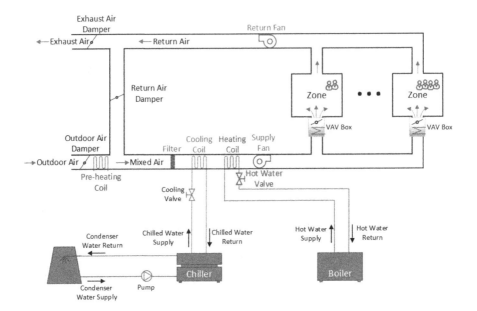

FIGURE 1.6
A typical HVAC system with gas-fired boiler and water-cooled chiller in a large commercial building.

inside the zone. Basically, VAV boxes maintain the desired comfort, usually the room temperature, inside the zones. Part of the combined air leaving the zones is recirculated and brought back into system. The amount of air recirculated into and exhausted out of the system are controlled by return air dampers and exhaust air dampers, respectively. At the same time, outdoor air entering the ductwork is mixed with the recirculated/return air. The amount of outdoor air is controlled by outdoor air dampers. In cold weather conditions, pre-heating coils are used to heat up the cold outdoor air to avoid freezing of the cooling coils further down the duct.

Once the mixed air passes through the filter, it is passed through the cooling coil where the temperature of the air is dropped to 55 °F (12.78 °C). Reducing the temperature to such a low value lowers the humidity ratio, which is the amount of water vapor present in the air. If many zones are demanding hot air, the air can be heated up by turning on the heating coils for better comfort and control. This way, the reheating at the VAV boxes turns on only when it is needed. The supply fan pushes the air to the VAV boxes and the entire process repeats. The supply fan is used to deliver air to the VAV boxes and then into the zones while the return fan is responsible for sucking the air out of the zones. The coils, fans, and dampers are considered as parts of the AHU. There are other components of an air handling system which are not presented here. For example, humidifier can be used inside the ductwork in dry climates. Instead of using a pre-heating coil, a heat exchanger can be used to transfer heat from exhaust air to outdoor air. Details of an AHU-VAV system can be found in the book [7].

Chillers supply chilled water to the cooling coils in the AHU and other equipment [8, Chapters 1, 2, and 3]. Supply air at the AHU extracts cooling energy from the chilled water. Valves and pumps in the chilled water loop are modulated to maintain the desired temperature of the air leaving the cooling coils. After the transfer of energy, the water returning from the cooling coil drops in temperature. The warm water enters back into the chiller, which rejects the heat of warm water and supplies chilled water to the AHU. The process repeats. The hot water (condenser water supply) from the chiller (usually many chillers) enters the cooling tower in which water is sprinkled from the top and the heat is rejected into the environment using fans. Pumps are used to push water throughout the entire system. Cold water from the cooling tower returns to the chiller. In a nutshell, chiller is an intermediate heat exchanger, which dissipates heat to the cooling towers [8, Chapters 9, 10, and 12]. However, in an air-cooled chiller, the heat is directly dissipated into the air and there is no need of cooling towers.

Boilers are used to heat up the water that is being supplied to the heating and pre-heating coils. The heating coils can be present at the AHU or at the VAV boxes. Although the figure represents a gas-fired boiler, other types of boilers can be substituted. In a gas-fired boiler, the gas or fuel is fired to heat up the water or generate steam, which further exchanges the heat with its hot water loop. Detailed functioning of chillers, cooling towers, and boilers can be found in the handbook [1].

1.3.2.1 AHU Systems

As noted above, there can be several configurations with many optional components based on the project requirements and constraints. Figure 1.7 highlights the configuration of AHU systems. Similarities and differences between the systems are summarized below:

CAV vs. VAV

As the name suggests, CAV (constant-air-volume) systems supply a constant amount of air throughout the system. This is achieved by using a constant supply fan without any dampers in the duct. However, VAV systems vary the supply of air to meet the demand by changing the supply fan speed or damper position(s) at the VAV boxes. The configuration shown in Figure 1.6 is an example of a VAV system. CAV systems are simpler than the VAV systems and they often contain reheating options. However, VAV systems offer better

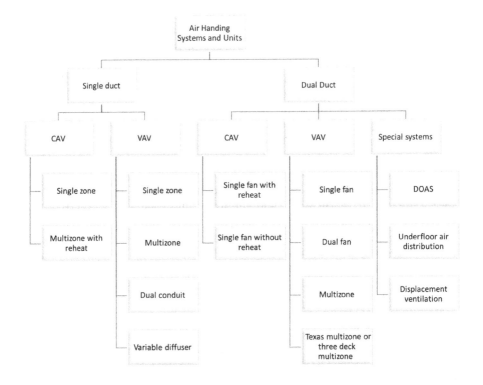

FIGURE 1.7
Types and configurations of air handling units and systems.

comfort and efficient control. Similar to the variations in the air handing units, there are several other configurations of the VAV and CAV boxes:

- *Fan powered series:* Extracts a portion of return or plenum air using a fan (acting as a booster for high air flow rate) that is in series with the discharged air.

- *Fan powered parallel:* Extracts a portion of return or plenum air using a fan (acting as a booster for high air flow rate) that is in parallel with the discharged air.

- *By pass:* Maintains the total air flow rate to a constant value, but bypasses a certain amount of air flow rate to the zone according to its control signal.

- *Induction:* Extracts warm air from the plenum and supplies to the zone without using a fan or blower.

Single Duct vs. Dual Duct

Single duct systems use one single duct for supplying conditioned air to the spaces or terminal units while dual duct systems use two separate ducts, one for heating referred to as hot deck and the other for cooling referred to as cold deck. In dual duct systems, the air is mixed at the terminal, closer to the zone or spaces, to modulate the desired temperature. While single duct systems have lower installation cost, dual duct systems are preferred in buildings that demands high ventilation requirement and possesses the zones with high variations in their loads. Typically, only one fan exists in the single duct configuration. Similarly, there can be two fans in a dual duct system. However, it is also possible to have one fan for both the ducts before the duct splits into two.

Single-zone vs. Multi-zone

Single-zone systems have only one single area controlled by the entire system in a same fashion. Single area could include multiple rooms or even the entire building. In single-zone, the system considers the entire space as one controlled area although there can be multiple components (sensors, diffusers, etc.) in the area. Auditorium is an example of large single-zone space. In contrast, in a multi-zone system, one system can control multiple zones independently. In a multi-zone system, one zone can be in heating mode while the other zone can be in a cooling mode. The example at the beginning of Section 1.3.2 corresponds to a multi-zone system.

Special Systems

Dedicated outdoor air systems (DOAS) are used to supply only fresh outdoor air into the spaces. The air leaving the spaces are exhausted into the environment, i.e., there is no return or recirculated air in the system. Displacement ventilation systems deliver air at lower velocity at low level inside a zone, which displaces the old air. The displaced old air is extracted at higher levels (e.g., ceiling) in the space because of buoyancy effects.

The main difference between underfloor air distribution system and other systems are that it uses a plenum under the floor and consider the thermal stratification during the design, which affects the supply air temperatures and the air distribution in a zone.

1.3.2.2   In-room Terminal Units

In addition to VAV and CAV boxes in zones, there are other terminal units, which are used to provide both sensible and latent heating/cooling to the end-space. They are called in-room terminal units. These systems do not rely fully on a central air distribution or other air distribution system. A few such systems are the following:

- Fan-coils

- Blower coils

- Unit ventilators

- Chilled beams

- Radiant panels

- Packaged terminal air conditioning (PTAC)

1.3.2.3   Heat Pumps

Figure 1.8 shows the configuration of a common residential HVAC system consisting of a heat pump with indoor and outdoor components. The indoor components of the heat pump include an expansion device, evaporator, and fan, whereas the outdoor components include a compressor, condenser, and fan.

The indoor fan (blower) pushes air though the evaporator coil(s), which conditions the supply air by rejecting heat and reducing the supply air temperature. The refrigerant in the evaporator coil is converted into gas due

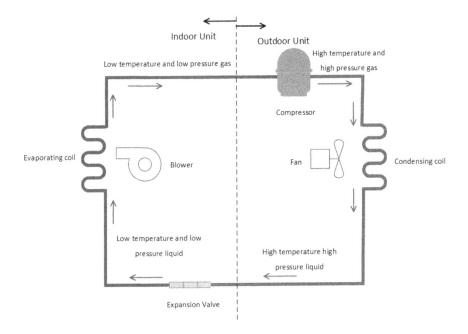

FIGURE 1.8
Configuration of a heat pump HVAC system used in residential buildings consisting of an indoor unit and an outdoor unit.

to the heat transfer between the supply air and refrigerant. Then, the hot refrigerant enters the compressor where its pressure is increased, resulting in a high-temperature refrigerant. The refrigerant, which is now in gaseous form, passes through the condensing coil(s) in the condenser. The outdoor fan blows air across the condensing coil resulting in the transfer of heat from the hot refrigerant to the outdoor air. During this process, the refrigerant temperature drops and the refrigerant is converted from gaseous to liquid state while maintaining a high pressure. The liquid refrigerant is then passed through an expansion valve, which decreases the refrigerant pressure and decreases the temperature further. The cooled refrigerant goes through the evaporator coil and the entire process is repeated. Although a typical configuration of HVAC systems in cooling mode is shown, the configuration can be reversed using reversing valves to provide heating. In heating mode, the condensing coil functions as an evaporating coil and the evaporating coil functions as a condensing coil. Gas furnaces and electric-resistance heaters are also commonly used for heating purposes.

1.3.2.4 VRF (Variable Refrigerant Flow) Systems

VRF systems are built upon direct expansion technology. They use a reverse Rankine vapor compression cycle and thus have many common components (e.g., compressor, evaporator, condenser, and expansion) when compared with a heat pump or chiller. VRF systems allow variable flow of refrigerant (as a primary medium for heat exchange) and simultaneous heating and cooling throughout the system. This enables efficient and precise control inside individual zones. A high-level diagram of a VRV system with simultaneous heating and cooling in multiple zones is shown in Figure 1.9. As shown in the figure, VRF systems has three main components: Outdoor units (ODU), indoor units (IDU), and heat recovery units (HRU).

IDU compares the room temperature against its set point to determine its mode (heating, cooling, ventilation, etc.) and sends a request to the ODU. Based on the requests from each IDU, ODU determines its mode of operation and delivers cold liquid and/or hot gas to the HRUs. The functioning of an ODU/IDU system is very similar to a heat pump with an exception that ODU also modulates the flow of refrigerant allowing simultaneous heating and cooling. The HRUs are responsible for forwarding the refrigerant to an individual indoor unit based on its original request from the indoor unit. Outdoor units can be used as a replacement unit for both air-cooled or water-cooled chillers in some scenarios such as retrofit projects.

While the VRF equipment cost higher than most other HVAC systems initially, a few advantages of VRF systems are:

- High efficiency and comfort

- Quieter operation

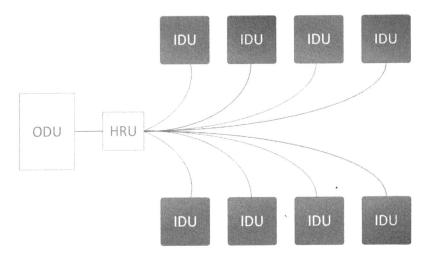

FIGURE 1.9
Schematic of a VRF HVAC system with heat recovery capabilities indicated by zones in heating and cooling modes at the same time.

- Smaller footprint because of piping and equipment sizing

- Lower installation cost during configuration/integration phase

- Lower operating and maintenance cost in certain conditions

While VRF systems have some advantages, it has not gained much traction in the US. As compared to other (or traditional) HVAC systems, a few disadvantages of VRF systems are controls, lack of flexibility to program the controllers at lowest levels, high capital cost, proprietary systems, inefficient usability and insufficient tools for debugging purposes, loss of efficiency when integrated into/with other systems, and less equipment/system life duration, and most importantly, the inflammable refrigerants, which can lead to potential safety and property-related issues such as refrigerants leakage, compressor failure, etc.

1.4 Control System Architecture

The control system structure in a large commercial building is shown in Figure 1.10. This architecture is applicable to several systems inside a building with general modifications. The common architecture in buildings is hierarchical in nature, the number of hierarchies is dependent on the

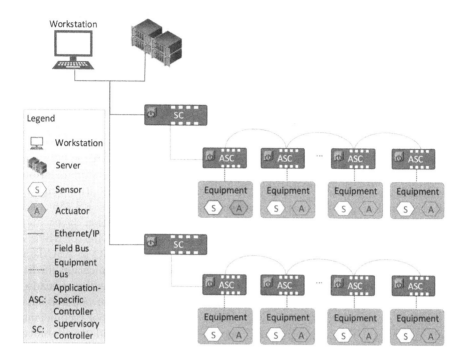

FIGURE 1.10
High-level system architecture in a large building.

complexity and type of building. In this architecture, the physical devices are controlled by application-specific controllers, which can be pneumatic, embedded, or a separate plug-in controller sending a hard-wired signal to the actual physical devices and equipment. Sometimes, ASCs (application specific controllers) are also called local controllers as they have sufficient memory, computational power, and logic to operate the devices and equipment in an isolated environment, i.e., ASCs are not relying on external signals for their fundamental operations. In a VAV box, the VAV controller processes the data obtained from temperature and air flow rate sensors to control its actuator, i.e., damper position. Although the lighting systems do not have as complex architecture as an HVAC system, a few lighting fixtures are getting equipped with temperature and occupancy sensors to control the lighting intensity at the local level. Similar to HVAC zones, lighting control systems also have zones in which a few light bulbs are controlled together. Lighting zone can be a single fixture or a collection of fixtures for a single room or a collection of rooms. A lighting zone does not have to be the same as an HVAC zone. One of the main reasons for zoning is to reduce the communication traffic and better organization of operations. It is possible that the lighting control system has the capability to control each light bulb independently.

Communication from embedded controllers to physical devices is very fast because of many reasons: direct connection, minimal protocol overhead, high baud rate, and short distance between the controller and device/equipment. However, when sensors and controllers are not part of an entire embedded system, the overall communication in the system can be slower. For example, use of thermostats and Wi-Fi sensors in the system can lead to much slower communication speed as compared to embedded systems. The physical devices are connected to each other in a daisy chain to enable communication between each other. The choice of communication protocols and communication physical layer (i.e., cable) depends on the type of system, control vendor, and the building configuration. In an AHU-VAV system, RS-485 and BACnet^{TM} [2] are commonly used as physical communication layer and communication protocols, respectively. In VRF systems, an RS-485 (or RS-485 type) cable and proprietary protocols are typically found. ASC-to-ASC is a slow communication channel because of the choice of communication media and protocols. RS-485 is normally chosen as a communication medium because of (1) its low purchase cost, (2) its low installation cost as other departments (e.g., IT) are not involved, (3) the knowledge and availability of technicians that understand the communication mechanism, and (4) wide availability of control products and external devices/sensors/equipment that support the RS-485 connections. Daisy chain is a cost-effective option because of the reduced number of ports and reduced length of cable regardless of the communication protocol and cable type. However, the configuration is very sensitive to the failures. Failure in a single device or the cable at any single point in the chain can stop communication between the rest of the devices. Star and ring topologies are available to increase the reliability, but these topologies increase the cost at the same time [6]. Therefore, an appropriate network configuration is chosen as per specific project requirements and constraints.

Supervisory controllers are connected to the application specific controllers using the same mechanism used between ASCs. Supervisory controllers are used to coordinate among the ASCs and different parts of the building systems to ensure that the entire system or sub-system is working as intended. In this process, the SCs collect and store data from other devices. SCs also transfer and report data to local severs or the cloud, which can be used for visualization and monitoring purposes. Because of the frequency of data transfers and large amount of data involvement, communication methods (e.g., TCP/IP, BACnet/IP or BACnet/Ethernet) with high bandwidth are used. Supervisory controllers also possess high computational power, and thus they are used to implement a system-level control algorithm or execute bulk operations. System-level logic improves the overall functioning of systems. As an example, the temperature set points of certain rooms can be lowered automatically during nighttime in a commercial building to reduce the overall energy consumption when no occupants are present. Details and implementation of a BAS architecture in commercial buildings are provided in articles [4, 5].

In a fire safety system, the role of supervisory controllers is minimal as they are used to obtain data from individual systems only and generate an alarm. Individual systems include smoke detectors, fire/flame detectors, water flow sensors, and manual switches. The alarm information is communicated to building owners, fire department, and occupants. Visual messaging displays in addition to standard notification appliances (horns, speakers, etc.) are used to relay the information. In this system, there is no control at the supervisory level. Furthermore, the local control is independent, simple, and mostly pneumatic. Similar to HVAC and lighting zones, a fire system can be divided into fire zones. It is important to note that the lighting, HVAC, and fire zones do not need to be the same as their purposes are different.

The choice of SCs is highly dependent on the building type and its principal activity/usage. For aforementioned reasons, it is easy to justify the presence of SCs in large commercial buildings. However, in a small-sized building such as business office or small retail store, a supervisory controller may not be needed because of its additional cost and security concerns. Moreover, the control opportunity and use cases of an SC may be very limited in small to mid-size building. Therefore, the benefits of SC in a small building are not justified against the additional cost and concerns. In contrary, if the building is part of a big-chain retailer irrespective of its size, supervisory controller can be utilized to access the data and manage a portfolio of buildings from a central location. Smart thermostat is an example of supervisory controller, which also possesses some of the ASC capabilities, e.g., local control. Many smart thermostats have embedded temperature and occupancy sensors that are used to make lighting and HVAC decisions. Furthermore, they integrate with home safety and security systems as well. Details of architecture and control-related products in residential buildings can be found in the report [3].

Other less-prevalent network architecture in building systems include star topology and ring topology. In a star topology, one ASC (or a limited number of daisy chained ASCs) is directly connected to the SC or the workstation. Start topology is expensive because of additional wiring but most reliable and robust as the wiring issues in one link do not affect the other controller. In a ring topology, the controllers are connected in a daisy-chain but the last controller is connected back to the first ASC or SC making a full ring. Ring topology is more reliable than a standard daisy chain but less reliable than the star topology from communication point of view.

In single-family homes, a simplified form of network architecture exists without any server or workstation. For smart systems, the controllers and the devices are connected directly to a cloud server through the Internet. Desktop applications or smart phone apps are made available to access or make changes to the system through the cloud. Add-on sensors are either connected directly to the cloud using a flat architecture or to the controller/thermostat that is ultimately sending information to the cloud server. Building systems in homes are typically isolated from each other, i.e., they don't share data between each other, although there may be a single application/app to visualize and control all the systems from one platform.

1.5   Moving toward the Future: Smart/Intelligent Buildings

Buildings consist of heterogeneous components and equipment, diversified controls, and multiple communication protocols. Basically, each building is a complex structure with a system of systems. The question is: how do we make the buildings "better"? "Smart/Intelligence" comes into play to answer this question. Smart buildings and intelligent buildings are used interchangeably in the book. There are multiple definitions of smart buildings without any consensus or clear characteristics. It is a relative term indicating the state of a building as compared to the state-of-art in most existing buildings. Intelligent buildings—as compared to most buildings—produce the following outcomes:

1. Improve the health/life and productivity of occupants;

2. Impact the environment positively;

3. Benefit society and other stakeholders; and

4. Maintain or advance safety, security and privacy at appropriate levels.

Life and productivity of occupants can be improved by providing better thermal comfort, IAQ, and creating an environment with tools that help people achieve their best performance. Allowing occupants to personalize their indoor environment in a convenient fashion, when possible, can yield higher productivity and better comfort, e.g., changing the temperature setpoint or the ventilation rate from a smart device such as smartphone.

Environmental benefits include lower energy usage, less carbon emissions, and reduced wastage of resources such as clean water and fuel. These benefits can be realized by replacing the existing equipment with high-efficiency equipment or upgrading the operations. The benefits also promote sustainability while having both environmental and economic impacts on other stakeholders including building owners, facility operators, etc. The use cases and the corresponding benefits depend on the type and priority of stakeholders involved in the situation. For example, if the components of building systems are integrated together through a unified platform or dashboard, building operators or facility managers can access the single system and easily detect and diagnose the problem areas such as faults, equipment malfunctioning, and unauthorized access. Details on the stakeholders are provided in Chapter 7.

Some of the aforementioned use cases not only affect an individual but also the society as a whole. Another example of societal impact is when multiple buildings, owned by different individuals or organizations, are working together to increase their power consumption in case of surplus renewable energy and vice-versa. It is also important that all such advancements consider the safety, security, and privacy of occupants and other stakeholders, including data security. For instance, the security of occupants can be enhanced by designing a system that automatically notifies security guards, calls police, and recommends next steps to occupants dynamically in case of criminal attack or fire in the building. However, if the potential solution requires installing a video camera in everyone's office, it will raise privacy concerns and may not be a viable solution in many situations.

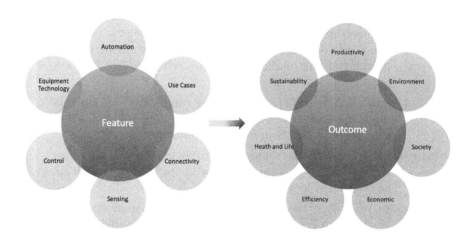

FIGURE 1.11
Mapping from the enabling features of smart/intelligent buildings to the outcomes and characteristics of intelligent buildings.

A few enabling features of an intelligent building are connectivity, automation, instrumentation, efficient equipment, smart control, and actionable use cases that improve over time. A high-level overview of an intelligent building's features and outcomes are shown in Figure 1.11. In a nutshell, intelligent buildings empower people, encourage sustainability, and improve the overall health and productivity of people, society, and the environment. It does not mean that every building needs to possess every feature and outcome to be considered as a smart building. It is a relative term and the use of "smart" is highly dependent on the context.

Key Takeaways: A Few Points to Remember

1. Buildings are the largest energy (40%) and electricity (70%) consumers in the US with HVAC as the largest contributor (40%).

2. Major systems in buildings are HVAC, power, fire safety, security, lighting, refrigeration, electronics, domestic hot water, plug-loads, and in-building transportation.

3. Vast amount of data (energy usage, activity, regions, floor area, age, fuel and system types, etc.) over a few decades for commercial and residential buildings is available publicly, which can be sliced and diced in many ways to extract meaningful information for both technical and business purposes.

4. AHU-VAV and RTU are the most common HVAC systems used in commercial buildings. Heat pumps are commonly used HVAC systems in residential buildings.

5. Most common control architecture in commercial buildings is hierarchical with controllers at multiple layers executing algorithms at local and supervisory levels.

6. In residences, the control architecture is flat or star type where controllers/devices/equipment are connected directly to the Internet with minimal interaction between the sub-systems.

7. A building ecosystem is complex in nature consisting of heterogeneous components and equipment, diversified controls, multiple communication protocols, and communication media.

8. Smart/intelligent buildings impact the health, quality of life, productivity, environment, and society in a positive manner while maintaining/advancing the stakeholders safety, security, and privacy.

Bibliography

[1] ASHRAE. *ASHRAE Handbook 2016: HVAC Systems and Equipment: SI Edition.* ASHRAE Handbook of Heating, Ventilating and Air-Conditioning Systems and Equipment SI. ASHRAE, 2016.

[2] ASHRAE. BACnet–A Data Communication Protocol for Building Automation and Control Networks. `http://www.bacnet.org/`, 2016. Accessed: 2017-06-25.

[3] M.C. Baechler C.E. Mertzger, S. Goyal. Review of residential comfort control products and opportunities. Technical report, Pacific Northwest National Laboratory, Dec 2017.

[4] Siddharth Goyal, Weimin Wang, and Michael R Brambley. An agent-based test bed for building controls. In *American Control Conference (ACC), 2016*, pages 1464–1471. IEEE, 2016.

[5] Siddharth Goyal, Weimin Wang, and Michael R Brambley. Design and implementation of a test bed for building controls. *Building Services Engineering Research and Technology*, 40(6):758–766, 2019.

[6] Johnson Controls Inc. Metasys IP Networks for BACnet/IP Controllers Technical Bulletin. `https://cgproducts.johnsoncontrols.com/MET_PDF/12012458.pdf`, Dec 2017. Accessed: 2019-12-30.

[7] L.A. Meyer. *Variable Air Volume Systems.* Indoor Environment Technician's Library. Lama Books, 1998.

[8] H.W. Stanford. *HVAC Water Chillers and Cooling Towers: Fundamentals, Application, and Operation.* Mechanical Engineering. CRC Press, 2003.

[9] US Department of Energy. How Energy-Efficient Light Bulbs Compare with Traditional Incandescents. `https://www.energy.gov/energysaver/save-electricity-and-fuel/lighting-choices-save-you-money/how-energy-efficient-light`, 2010. Accessed: 2020-08-02.

[10] US EIA-Department of Energy. 2012 CBECS detailed tables, 2016.

[11] US EIA-Department of Energy. 2015 RECS detailed tables, 2018.

[12] U.S. Energy Information Administration (EIA). Electric power annual 2017. Technical report, United States Energy Information Administration, Dec 2018. Accessed: 2019-06-30.

[13] U.S. Energy Information Administration (EIA). *Annual Energy Outlook 2019*. DIANE Publishing, Jan 2019. Accessed: 2019-06-30.

[14] U.S. Environment Protection Agency (EPA). Indoor air quality. `https://www.epa.gov/report-environment/indoor-air-quality#note1`. Accessed: 2019-06-30.

[15] U.S. Environment Protection Agency (EPA). Sources of greenhouse gas emissions. `https://www.epa.gov/ghgemissions/sources-greenhouse-gas-emissions`. Accessed: 2019-06-30.

2

Sensing

> "If you cannot measure it, you cannot improve it."
>
> — Lord Kelvin

Sensor is an electronic device that produces electrical, optical, or digital data derived from a physical condition or an event. Data produced from sensors is electronically transformed, by another device, into information (output) that is leveraged in the decision-making processes adopted by other devices or a person. Sensing is a technique used by sensors to measure a physical property or an event in the environment. Sensors are prevalent in the building systems as they generate vast amount of useful data every day. A large commercial building with a BAS generates millions of measurements (i.e., sensors' data) each day when sampled at 1-minute frequency. Use of sensors have been increasing over the past few years in the building industry because of their decreasing cost and their increasing utilization in providing value to their customers. This had been possible because of (1) recent development of new technologies in both hardware and software, improvements in mass production facilities, and (2) use of same sensors across multiple industries. As the accuracy, reliability, and affordability of sensors have been progressively increasing over the past, sensors have been playing more active role in the modern control systems in buildings. Instead of sensors being used only for monitoring purposes, they are getting integrated actively into the building systems to improve the overall system performance.

2.1 Background and Overview

This section provides an overview of the characteristics and categories of sensors used in buildings.

2.1.1 Characteristics

- Cost: Cost is one of the most important factors in choosing a sensor in building systems. This includes not only the cost to purchase a sensor but also the cost to install, operate, and maintain the sensor, conditional to the type of stakeholders involved in the process. For example, a contractor installing a sensor focuses on the installation cost and perhaps the purchasing price, but may not consider the operational and maintenance costs.

• Connectivity: Connectivity defines the interfaces offered by a sensor to communicate data to other devices. Sensors can be connected to other devices (or sensors) wirelessly (Wi-Fi, Bluetooth, Zigbee, etc.) or through a hard-wire such as 0–5V analog output. Connectivity is important in buildings as sensors with wireless connections can reduce both installation and maintenance costs. Connectivity to the Internet enables monitoring and implementation of applications from a remote location. On the other hand, hard-wired sensors may have higher reliability and less cybersecurity concerns.

• Sampling rate: The rate at which a sensor generates data to measure the change is called sampling rate. Sampling rate is essential for determining the control time step, which is the time period between two consecutive control actions. Sensors in buildings have sufficient sampling rate for existing applications such as temperature set point control, damper control, and lighting control. However, there are certain advanced applications (e.g., compressor or fan control for frequency regulation) that require higher sampling rate than what is normally present in buildings.

• Power source: There are a few options for supplying power to sensors: (1) battery operated, (2) a separate plug-in power source, and (3) power from a controller/equipment embedded in the equipment. Power source is a factor considered during installation and maintenance phases in buildings.

• Interoperability: Interoperability of sensors is the ability to share, understand, and exchange information from one sensor to another sensor, device or equipment in a well-understood fashion/form. Open communication protocols are essential to achieve high interoperability. Amount of data available to other devices, typically through an application programming interface (API), also contributes to the interoperability.

• Location: Location of sensors in buildings is as important as other factors. The sensors can be installed outdoors and indoors in spaces, ducts, plenums, pipes, devices, and equipment. Installation location can be constrained by the type and characteristics of sensors. For example, water resistant or waterproof sensors is a necessary requirement to install the sensors outdoors for certain applications.

• Accuracy and resolution: Accuracy is an absolute measure of precision to quantify the correctness of observed quantity. $\pm 1\%$ accuracy of a sensor means that the measured values obtained from the sensor are within 1% of the true values. Resolution is the smallest increment in the measurements that a sensor can display or record.

- Size, volume, and weight: In general, size, volume, and weight are major attributes of a sensor. In building systems, size is at utmost importance in various devices or controllers because the size of sensors can affect the form factor or overall size of the devices, e.g., thermostats or remote temperature sensor in a house. It the sensor is a stand-alone device, these attributes contribute to the aesthetics and customer perception.

- Smartness: Smartness cannot be easily quantified as there is no standard definition on the smartness of sensors. In the building systems, smartness relates to the features, which improve the overall working and use of the sensor. The smart features are usually embedded in the sensor or a stand-alone sensing device. These features could include, cleansing and filtering of data to provide precise readings in uncertain conditions, storing the measurement internally for backup reasons, user-interface for easy configuration, and plug and play connectivity with minimal additional effort.

- Other: Several other technical characteristics of a sensor are errors (drift and bias), linearity, sensitivity, range, reliability, and repeatability. Details on these characteristics can be found in [7, 8].

2.1.2 Analog vs. Digital

In buildings, sensors are categorized in two major groups, digital and analog. Digital sensors, also called as binary sensors, produce binary outputs with only two possibilities (e.g., 0 or 1, true or false, on or off) while analog sensors produce continuous outputs in a certain range. Temperature, pressure, air flow rate and other sensors described above are analog sensors. A couple of examples of binary sensors are the PIR (Passive Infrared) occupancy sensors to detect presence and the sensors that detect the run status of fans, pumps, and motors, i.e., detecting whether the equipment/device are on or off.

2.2 Virtual Sensors

Virtual sensor is a unit that infers the property of an element using data from other sensors or other properties of the same element or different elements. Virtual sensor can be just a software module or a physical device running software modules. Virtual sensors use modeling, estimation, and system identification techniques. The use of virtual sensors is obvious when the actual sensor is not available. However, if a real sensor is available, one may ask why we need a virtual sensor. The following are the other reasons or the scenarios justifying the case:

- *Validation:* To identify the faulty sensors using validation and verification approaches, e.g., checking inconsistencies in mixed air and supply air temperatures [9].

- *Purchasing cost:* To reduce the cost of installed system because the actual sensors are very expensive, e.g., BTU meters and flow stations are expensive and not a viable solution for every system.

- *Calibration:* To detect bias and drift errors in the sensors and update the calibration factor in control algorithms.

- *Maintenance and operation:* To avoid or reduce the operating and maintenance cost of existing sensors, e.g., battery operated sensors with low battery life when data is frequently transferred over Wi-Fi.

- *Easy updates and installation:* Deploying and updating a software module is much easier than replacing a physical sensor, especially when the sensor or the device holding the sensor is connected to the Internet.

- *Sampling rate and resolution improvement:* To improve the characteristics of an existing sensor, e.g., providing ancillary services to the grid at 4-sec intervals while most BAS cannot support data communication on 4-sec intervals.

- *Noise filtration:* To improve the functioning and accuracy of existing sensor by incorporating latest tools and technologies, e.g., use of smart-phones and existing occupancy sensor to count the number of people in a room.

- *Backup:* To use virtual sensor as a backup if the original sensor is temporarily unavailable, typically in critical facilities. If the communication signal is lost for long time or if the sensor stops working, a virtual sensor can be utilized temporarily as a backup solution.

- *Physical limitation:* Virtual sensors are preferred methods because of sensor location and other physical constraints, e.g., retrofit projects in museums or historical architectural buildings may require minimum installation changes to preserve the originality.

- *Others:* To overcome the computational and cybersecurity limitations of existing sensors.

2.2.1 Analytical

Analytical virtual sensors use a pre-defined mathematical model to estimate the quantities of interest. It is used in scenarios when the relationship between the variables is well-understood and the behavior can be modeled with available tools and techniques, normally using first principles. Estimating the number of people from CO_2 measurements is an analytical virtual sensor. However, there are some scenarios in which non-modeling techniques are used to estimate the number of people. Fault detection and diagnostics (FDD) based on energy-balance principles and simple logics are applications of analytical virtual sensing [9].

2.2.2 Empirical

Empirical virtual sensors use a data-driven approach for estimation. In this method, current and historical data are used in a statistics-based method to estimate an event or physical property in the environment. The least squares approach [13, 10] is a simple empirical method for virtual sensing. Other methods include neural networks and machine learning (ML). Detecting a person's activity from the sensors present in smart phones using a ML technique is an example of empirical virtual sensing. A combination of empirical and virtual sensing can also be used in building applications.

2.3 Sensors Types and Availability in Buildings

2.3.1 Temperature Sensors

Temperature sensor is one of the most frequently used sensors in buildings because of its low cost and relevance in many applications. They are used to measure the temperature of substances in the building systems such as air, water, refrigerant, steam, gas, and liquids. In Figure 1.6, the temperature sensors measure zone air temperature, supply air temperature, return air temperature, mixed air temperature, outdoor air temperature, chilled water temperatures (supply and return), condenser water temperatures (supply and return), and hot water temperatures (supply and return) at boilers. A few locations of temperature sensors include ducts, plenums, zones, tanks, freezer, pipes, outdoors, and under the floors. To measure the temperature of water, refrigerant, or other liquid, the temperature sensors are wrapped around the pipes while the sensors are physically present in ducts/decks or plenums to measure the air temperature. Because of their low cost, temperature sensors have started to become part of several building systems such as lighting, irrigation, and cooking.

Location of temperature sensors is important as it can influence the measurements significantly. Temperature sensor either close to window being exposed to the sun or close to the diffusers is not a true reflection of zone temperature. Stratification can cause up to 6 °F (3.33 °C) different in temperature in a single zone [12]. In addition to the location, the sensor type defines the accuracy, cost, and reliability of a sensor. Thermocouples, thermistors, and resistance-temperature devices (RTD) are three major types of temperature sensors in buildings. Their pros and cons are briefly discussed in the report [5].

2.3.2 Pressure Sensors

Pressure sensors in buildings measure static pressure, dynamic pressure, and differential pressure across two points in a loop. Static and dynamic pressures are measured for the locations with fans, i.e., ducts. The pressures of return air, supply air, mixed air, exhaust air, and discharge air are obtained using such sensors. Boiler pressure and building static pressure are also measured in buildings. In healthcare facilities, it is required to have a positive pressure to avoid contamination via infiltration, i.e., outdoor air entering the building through cracks or leaks. Differential pressure sensors measure differential pressure across hot water loop (supply and return), chilled water loop (supply and return), condenser water loop (supply and return), refrigerant loops in heat pumps and VRF systems. Differential pressure sensors are deployed across fans and filters in some cases. Differential sensors are utilized in systems with variable speed fans or pumps.

2.3.3 Flow Rate Sensors

Flow rate sensors measure the air flow rate and liquid flow rate in the building systems. Air flow rates are obtained for mixed air, return air, supply air, exhaust air, discharge air, and fume hood air. Water or refrigerant flow rate sensors are used for measuring flow rates for chilled water, hot water, condenser water, and refrigerants. Flow stations or flow meters have high accuracy but they are quite expensive as getting an accurate air flow rate in duct or any other enclosed pathway requires measuring velocity at different points in the duct. Therefore, a virtual sensor is used as an alternate in which pressure sensor or a velocity sensor is used first to measure the value at one point in the duct, and later combined with the duct area to estimate the air flow rate. Sometimes accurate flow meters at terminal units are temporarily used for calibration and validation purposes to ensure that the zones are getting enough air flow rates as per design specifications and building codes.

2.3.4 Speed Sensor

These sensors measure the speed of motors in pumps and fans. Variable frequency drives are available to report the measurements. The speed of fans and motors are automatically controlled to maintain a predefined differential pressure across multiple loops. Although the speed is controlled internally, the measurements are available for diagnosis and monitoring purposes. Wind speed of outdoor air is also measured using speed sensors.

2.3.5 Electrical Sensors

Voltage, current, frequency, and power sensors (also called as power meters) are considered in this category. These sensors measure input and output values (voltages, current, frequency, active/reactive power) in both AC and DC circuits. They can be available in electrical equipment in various building systems: battery or solar panel in power systems, lighting control panels, heat exchanger and variable speed drives in HVAC systems, and hot water systems. The measurements may be available to the users, but they are not fully utilized by the users today. Sometimes they are used for diagnosis and maintenance purposes to check whether the equipment is working. However, they can be used for active control and automatic FDD. One example is to modulate the power and frequency of equipment inside the building to provide grid services at different time scales; refer to article [6] for the details. This is an active research topic in the community. Several other sensors also generate low-voltage or low-current as their outputs, e.g., temperature sensor's generation of 0–5 V; such sensors are not classified in this category.

2.3.6 Humidity Sensors

Humidity sensors are used to measure absolute and relative humidity of supply air, return air, discharge air, exhaust air, zone air, and outdoor air. Although zone temperature is only controllable variable in most HVAC applications, some applications require controlling both temperature and humidity in a space to yield better comfort, higher efficiency, and tighter indoor environment. However, if the humidity sensors are not accurate, especially while using inexpensive humidity sensors, they can lead to inefficient control as shown in the study of economizer control [14]. Humidity sensors are also combined with temperature sensors to calculate dew points.

2.3.7 Light Sensors

There are three types of sensors used to (1) measure the intensity of light in a space, (2) detect the type/distribution of light, e.g., detection of natural light (vs. artificial light), and (3) measure the direction of lighting source. Use of different types of light sensors have been increasing over the past few years. However, lighting control is not progressing at fast rate today primarily because of development of highly-efficient lighting fixtures that last for a while. Therefore, the value proposition needs to be stronger than just energy savings to enable advanced lighting sensors and advanced lighting control systems.

2.3.8 Occupancy/Motion Sensors

Motion sensors are used to detect motion from any object in a space while occupancy sensors are used to detect people, i.e., presence/absence or the number of people. There is significant overlap between these sensors. PIR is the most common technology used to detect motion or a person's presence in an area. PIR sensors are also used to count the number of people entering through doorways. PIR sensors are very inexpensive (the least expensive occupancy sensor) and thus found in many devices and appliances, e.g., smart faucets, smart phones, thermostats, light fixtures, and security systems. Ultrasonic sensors detect motion through ultrasonic sound waves. As people generate heat, moisture, and contaminants at a certain rate, CO_2, temperature, and humidity sensors are used as virtual sensors to count the number of people.

Thermal imaging sensors count the number of people in a room through heat signatures. Image processing from camera(s) is another option to estimate the number of people. However, use of camera in every space is not only expensive but also raises privacy concerns. Other technologies to count the number of people include tracking cell phones, the connection of cellphones with Wi-Fi access points, the number of active users on desktops, and badge readers. A combination of these sensors can also estimate the number of people in a space.

2.3.9 Security Sensors

Cameras, PIR sensors, window sensors, door sensors, and garage sensors are used as security sensors in a home or commercial building. In homes, cameras are installed outdoors and indoors to detect and notify suspicious activities. In commercial buildings, cameras are continuously monitored to deter crime, minimize loss, mitigate risks, reduce false alarms, and manage dispersed facilities remotely. PIR sensors are used to detect motion during unexpected events, e.g., thief entering through backyard at midnight. Similarly, door, windows, and garage sensors are used to notify homeowners or security professions if they are kept open when they are not supposed to.

2.3.10 Air Quality Sensors

Air quality sensors are used to measure the quality of both indoor and outdoor air. These sensors include measurements of volatile organic compounds (VOC), CO (carbon monoxide), SO_2 (sulfur dioxide), CO_2 (carbon dioxide), particulate matter 10/2.5, and pollen. Portable sensor packages that combine multiple sensors (VOC, PM, CO_2, and CO level/concentration) are available to measure the IAQ. Photoionization detectors (PIDs), infrared carbon dioxide monitors, and metal oxide sensors (MOS) are available to detect multiple types of VOCs. Similar to weather websites [11, 1], commercial websites and products are available to measure the outdoor air quality and pollen level in real-time. Smoke detectors are present in the buildings as part of fire safety system. In many buildings, there are no or very limited air quality sensors. The most common air quality sensor installed in commercial buildings is CO_2 sensor. If a CO_2 sensor is present, it is normally used to control the ventilation rate, i.e., controlling the fresh outdoor air directly or indirectly by changing the outdoor air damper, total supply air flow rate, outdoor air ratio, or return air damper. This is called demand control ventilation (DCV) [4]. If other IAQ sensors are available in buildings, they are present to monitor, notify, detect major problems and notify the users through alarms in a manual fashion.

2.3.11 Outdoor Environmental Sensors

In addition to temperature, humidity, and air quality sensors mentioned earlier in this section, wind direction, solar radiance, sun angle, and rain sensors are a few other sensors used to measure the properties of an outdoor environment. Rain sensor is used to detect rain outdoors so that the irrigation system can be turned off or on accordingly. Sun angle sensors are used to determine the solar azimuth and zenith angles, which could be found in solar panels to maximize the power generation. Other sensors could be occasionally found in buildings.

2.3.12 Position Sensors

Valve and damper positions are acquired from position sensors. In Figure 1.6, position sensors measure the position of outdoor air damper, return air damper, mixed air damper, exhaust air damper, supply air damper, chilled water valve, hot water valve, and condenser water valve. Deploying an actual position sensor for these measurements is not inexpensive and thus a surrogate method is typically used. In this method, the voltage or current signal sent to damper as a command is considered as its true value. In short, the delay from controller to the actuator and the response from the actuator is ignored while estimating the damper position. However, real position sensors are commercially available if true feedback and accurate position of dampers/valves is needed for certain applications or projects.

2.3.13 Safety Sensors

These sensors are used to measure or monitor conditions that can cause physical damage to a person, building, property, or asset. Sensors in this group are heat sensors, flame sensors, CO/CO_2 sensors, fire sensors, water-level sensors, and refrigerant leak sensors. Heat, flame, and CO/CO_2 sensors detect potential fire and activate the fire suppression system automatically. Water-level sensors are typically installed in homes to detect water leakage to avoid property damage. When people are on a vacation away from home, such sensors can detect the excess water and notify the people in a timely fashion on their phone or via an email. A manual action or intervention is needed afterward. Similarly, refrigerant leak sensors detect leakage in VRF or heat pumps, which also demand a manual action afterward.

2.3.14 Meters or Usage Sensors

Meters calculate the consumption of thermal energy, power, water, steam, and gas in buildings. Water, electricity, and gas meters are found in almost every building so that the utilities can charge the building owners or facility managers accordingly. Although these meters are used primarily for billing purposes, new meters that can turn on and off the power supply remotely have been increasing in residences.

Advanced metering infrastructure (AMI) is a technology that allows two-way communication in real-time between utilities and buildings owners or operators through meters [15]. AMI is capable of combining data from all water, gas, and electricity meters at high-resolution. Although AMI requires initial investment to install the necessary hardware and software modules in a building, AMI offers several benefits to both utilities and building owners in terms of accuracy, convenience, operation, control, and service speed.

Thermal energy meters (also called as "BTU meters") calculate the load or thermal energy consumption across a loop. For example, BTU meters can be installed at individual VAV boxes to calculate the total energy consumption at individual zones. Because thermal energy meters are quite expensive, virtual sensors are used to estimate the thermal energy consumption. However, obtaining an accurate estimation of thermal energy is quite cumbersome because (1) several required sensors that are not typically present in buildings—SA temperature, humidity, and air flow rate sensors—are needed at the VAV boxes, and (2) the resolution and accuracy of the such sensors may not be adequate. Similarly, an application of water meters in commercial buildings with cooling tower is to detect leakage or calculate makeup water, which is the water lost during the condensation process. Usage of domestic water is also calculated using water meters. Flow sensors are also called flow meters. Enthalpy sensors are also available to be installed in buildings though they are not typically installed in buildings because of their limited benefits, high cost, and low accuracy. However, if installed, they can be found useful as part of economizer control in certain climate conditions where the energy from outdoor air can be significantly utilized.

2.3.15 Tracking Sensors

GPS (Global Positioning Systems) is normally used for tracking the location of an object or a person in this industry. The location can be used in several applications. Existing residential thermostats utilize the GPS to determine the room temperature set point and the start time of an HVAC system. The GPS helps the system determine the location of occupants through a mobile App in their smart devices. The resolution of GPS is not enough to determine the location of occupants inside a building. Therefore, other technologies such as Wi-Fi connections with access points, radio-frequency identified (RFID) tags, badge readers, and computer activities are used to track occupants inside the buildings depending on the applications. A combination of PID sensors is also used in grocery or retail stores to observe people behavior and control lighting.

2.3.16   Other Sensors

There are several other sensors that could be found in buildings depending on the applications and use cases. A few of them include velocity sensors, water conductivity sensors, torque sensors, outdoor grain sensors, and other acoustic sensors such as detection of people from walking sounds or microphones. Other examples of commonly used acoustic sensors are voice-recognition sensors and smart sensors in residential buildings such as Amazon Echo [2] and Apple HomeKit accessories [3].

As described earlier, building systems employ several types of sensors in multiple applications. These sensors utilize a wide range of technologies offering different characteristics and features. Figure 2.1 summarizes the sensing technologies and sensors used in the buildings. The list of technologies used in buildings sensors described before are infrared, ultrasonic, piezoelectric, resistive, conductive, capacitive, electromagnetic, LiDAR, sonic, image processing, thermal, GPS, optical, semiconductor, ionization, potentiometer, photoelectric, hydroscopic, metal-oxide, calorimetric, and voice recognition. Note that multiple technologies can be used in one or more sensors, e.g., GPS can be used in tracking sensors and occupancy counting sensors.

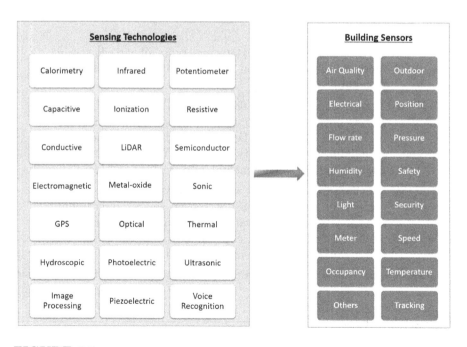

FIGURE 2.1

Types of sensors in building systems and the technologies associated with the sensors in alphabetical order.

"What we know is a drop, what we don't know is an ocean."

Newton, *Isaac*

Key Takeaways: A Few Points to Remember

1. Features and characteristics of a sensor, in general, include cost, connectivity, sampling rate, accuracy, resolution, error, linearity, range, reliability, repeatability, sensitivity, size, weight, volume, power source, interoperability, location or physical constraints, and smartness from user's perspective.

2. Virtual sensing is a way to estimate the value of physical property or an event from other sensors. Several applications and benefits of virtual sensors are data validation, noise filtration, sensor calibration, low purchasing and deployment cost, easy/quick updates, and temporary replacement of sensors during malfunctioning, maintenance, and operation stages.

3. Analytical and empirical are two types of virtual sensing methodologies based on model-driven and data-driven techniques, respectively.

4. Temperature, pressure, flow rate, fire safety, meters, and electric sensors are the most common sensors installed in buildings. A few other sensors that could be found in buildings are humidity, speed, occupancy, security, air quality, outdoor environmental, and tracking sensors.

5. Because of large number and types of sensors with varying complexity and trade-offs, a sensor should be carefully chosen. The most relevant characteristics of sensors for intelligent buildings are cost, connectivity, sampling rate, power source, accuracy, size, and smartness.

Bibliography

[1] Weather underground. `https://www.wunderground.com`. Accessed: 2020-11-30.

[2] Amazon. Amazon Echo & Alexa Devices. `https://www.amazon.com/smart-home-devices/b?ie=UTF8&node=9818047011`. Accessed: 2020-06-01.

[3] Apple. Apple Home and HomeKit. `https://www.apple.com/ios/home/`. Accessed: 2020-12-01.

[4] Michael Apte. A review of demand control ventilation. *Proceedings of Healthy Buildings*, 4, 01 2006.

[5] M.C. Baechler C.E. Mertzger, S. Goyal. Review of residential comfort control products and opportunities. Technical report, Pacific Northwest National Laboratory, Dec 2017.

[6] Siddharth Goyal, Weimin Wang, and Michael R Brambley. An agent-based test bed for building controls. In *American Control Conference (ACC), 2016*, pages 1464–1471. IEEE, 2016.

[7] Ricardo Gutierrez-Osuna. Intelligent sensor systems. `http://courses.cs.tamu.edu/rgutier/ceg499_s02/l2.pdf`. Accessed: 2020-08-02.

[8] Jim Irish. Lecture on Instrumentation Specifications, Ocean Instrumentation, Course 13.998. `https://ocw.mit.edu/courses/mechanical-engineering/2-693-principles-of-oceanographic-instrument-systems-sensors-and-measurements-13-998-spring-2004/readings/lec2_irish.pdf`. Accessed: 2020-08-02.

[9] S. Katipamula, W. Kim, R.G. Lutes, and R.M. Underhill. Rooftop unit embedded diagnostics: Automated fault detection and diagnostics (AFDD) development, field testing and validation. Technical report, Pacific Northwest National Laboratory, 2015.

[10] Charles L Lawson and Richard J Hanson. *Solving least squares problems.* SIAM, 1995.

[11] US Department of Commerce and NOAA. NWS JetStream – Climate, Aug 2019.

[12] James Piper. How To Fight Air Stratification in Conditioned Facilities. `https://www.facilitiesnet.com/hvac/article/How-To-Fight-Air-Stratification-in-Conditioned-Facilities--17985`, 2018. Accessed: 2020-08-02.

[13] M. Schomaker, C.R. Rao, H. Toutenburg, and C. Heumann. *Linear Models and Generalizations: Least Squares and Alternatives.* Springer Series in Statistics. Springer Berlin Heidelberg, 2007.

[14] Steven Taylor and C. Cheng. Economizer high limit controls and why enthalpy economizers don't work. *ASHRAE Journal*, 52:12–28, 11 2010.

[15] US Department of Energy. Advanced metering infrastructure and customer systems: Results from the smart grid investment grant program. Technical report, US Department of Energy, September 2016.

3

Modeling

Mathematical modeling, which is referred to as modeling in this book, is a way to capture and represent the system behavior/functioning using mathematical concepts and equations. Models describe the relationships between inputs and outputs using mathematics. Figure 3.1 shows an abstract representation of a mathematical model.

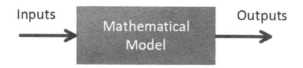

FIGURE 3.1
Abstract schematic of a model providing a relationship between inputs and outputs.

Inputs are fed to a model and the model generates outputs. It is expected that a model mimics a real-system so that the system behavior and the effects of inputs on the outputs can be studied and well-understood. If the same inputs from a real-system are provided to its corresponding model, the outputs from the model should be similar to the outputs from the real-system. The closer are the outputs from the model to the real-system, the higher is the prediction power or accuracy of the model. A simple example of a mathematical model is Newton's second law of motion, in which the input is the force and output is the acceleration of a constant mass object. There are three types of mathematical models: (1) White box, (2) Black box, and (3) Grey box, which are described next.

3.1   Modeling Approaches

3.1.1   White Box

White box approach uses first principle(s) or physics-based mathematical equations to create a model. The model uses prior knowledge of the systems and it is considered as "well-known" model from those insights. An example of white-box model is the Newton's second law of motion in which the output (acceleration) can be determined from a given input (force) and a fixed parameter (mass). Because of the nature of the model, it is very challenging to find an accurate white box model of a real-system that does not change much over time, especially in building systems. Therefore, they are not prevalent in real-time systems for control purposes. However, they can be found in simulation tools to understand the behavior of the systems. White box models offer several advantages as the model:

- Does not require any real-time or time-series data of the system. It may require fixed parameters as part of the model, e.g., fluid properties and wall material.

- Provides a clear visibility into the structure and the relationships between variables.

- Remains unchanged during uncertain conditions or remains the same with the availability of new data.

- Can be used to conduct theoretical analysis and quantify the performance such as error, stability, etc.

- Is applicable for a wide range of operations.

3.1.2   Black Box

As opposed to white box models, black box models rely on data to derive relationships between inputs and outputs using statistics. Linear regression and correlation methods are a couple of examples of black box modeling techniques. Neural networks and some of ML [13, 14, 35] methods are also considered in this category. Users have no or minimal visibility into the model, and thus called as "black box" model, sometimes referred to as purely data-driven model. In some cases, the technique is also automatically selected based on the data available from system. For instance, the algorithm can choose least squares first, and switches to neural-network (NN) later once more data becomes available. Similar to white box models, black box models have their own advantages and disadvantages.

Development of black box models is usually simple and fast as they don't require domain expertise, internal structure, or the system physics. However, developing an accurate black box model requires significant amount of training data, i.e., response and behavior of system for different inputs. The operating range of the model is usually dependent on the range of inputs provided during the training period. Higher the amount of data, the better is the model. The model may change as new data becomes available to the system. It is challenging to debug or interpret the results from such models as there is not explicit connection between the variables.

3.1.3 Grey Box

Interestingly, the advantages of white box models are the disadvantages of black box models in many situations and vice versa. Therefore, grey box models combine both white box and black box models to overcome their shortcomings and utilizing their strong features. Grey box is a hybrid modeling approach in which simplified physics-based equations are combined with simple data driven techniques to acquire only some of the model parameters/coefficients.

Similar to white box models, the structure of grey box models is known and thus the performance of a grey box model can be quantified as a function of its coefficients. Furthermore, it is easier to interpret results and debug problems with grey box models as compared to solving problems in black box models. However, the development of grey box models can be slow and time consuming because of training data and the identification, calibration, and validation processes involved in obtaining accurate coefficients/parameters. Although the operating range of a grey box model also depends on the training data, the model is expected to work better than the black box model outside the operating ranges because of first principles and physics-based equations used in the model. For all the above-mentioned reasons, grey box is most prevalent in control applications that are dependent on models.

3.2 Modeling Process

Figure 3.2 shows an overview of model development process. Modeling category is decided first including its type and features. Trade-offs in selecting models are discussed in Section 3.4. After the first stage of selection, the model may be reduced or approximated based on the requirements. The next step is to estimate or calibrate the model parameters. The calibrated model is validated against a pre-defined performance metric to finalize the model. Each step is explained in detail next.

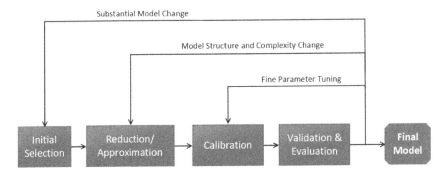

FIGURE 3.2
Model development process.

3.2.1   Model Types

In additional to high-level classification mentioned in the previous section, there are several other ways to characterize and identify a model. The types of models include linear models (or non-linear models), static models (or dynamic models), time-variant models (or time invariant models), continuous models (or discrete models), and deterministic models (or stochastic models). Linear models obey the principles of homogeneity and superposition. However, if a model contains any non-linear mathematical equation, it is considered as a non-linear model. Usually, the parameters of a model are fixed in time, e.g., mass of a block in Newton's second law of motion. If the parameters don't change over time, the model is time-invariant. If the parameters vary over time, the model is called time-variant model.

Static models show only the steady-state behavior between the variables independent of time. In contrast, dynamic models capture the transients and time-dependent behavior of a system as a function of time. Discrete models contain at least one discrete variable while continuous models contain continuous variables, e.g., temperature in a room is a continuous variable. In practice, many continuous models are converted into discrete models as the sensors generate data at a predefined interval. In deterministic models, the parameters, coefficients, or functions are represented uniquely in a deterministic fashion. However, stochastic models involve randomness with probabilities, i.e., the value of a variable can be calculated with certain probabilistic characteristic or confidence level. Stochastic models are least found in building systems because of their complexity. The model types and properties can be combined together to create a unique type of model. For instance, there can be a continuous linear time varying model or continuous non-linear time invariant dynamic model.

3.2.2 Order Reduction and Approximation

Model reduction, or model order reduction, is a set of techniques used to reduce the computational power or complexity of a model, typically by reducing the number of states, variables, or relationships. They are applicable to both online (real-time) and offline tasks. In simulations, model reduction enables faster simulation results so that the users can understand the system behavior quicker in preliminary stages. Once the system is well understood, a full order model can be implemented to confirm the findings. Model reduction becomes highly productive in the situations where the controller possesses limited memory and the control algorithm needs to have real-time predictions from the model. In real-time systems, controllers in building systems are not computationally powerful unless they are using distributed computation through the cloud. Balanced truncation, balanced realization, projection based, empirical gramians, POD (proper orthogonal decomposition), Krylov-subspace, and approximation methods are commonly used model-order reduction techniques [26, 17, 15]. Model-order reduction is not prevalent in building systems. In certain situations, model-order reduction techniques can be found in the video or image processing sensors as part of security systems. Similarly, approximation techniques can be implemented to reduce the model complexity. Approximation is conducted to improve the computational performance and obtain theoretical guarantees on the performance.

3.2.3 Calibration and Validation

Model calibration is the process of identifying the coefficients or parameters of a model within pre-specified constraints. Model calibration is done for a variety of models, particularly grey box models. In this process, the input data and corresponding response data from the system are collected over a period of time. Using the collected data, the model coefficients are tuned to optimize the model. Performance of the model is evaluated against a performance metric. A few performance metrics are maximum, absolute, or average deviation of predicted outputs from inputs. Sometimes, a wide range of impulse inputs are sent for sufficient persistency of excitation to improve the overall model performance in wide operating range. Linear regression, exhaustive searching, and optimization-based methods are a few available options to calibrate a model.

Once the model is calibrated, the next step is to validate the model. It is important to ensure that the model works well before it is deployed on real-systems. Therefore, the model is tested against a new set of data to confirm its performance. In this step, a sample of data is collected from the system during different times with multiple conditions/inputs. Response of the model is captured with the same inputs being sent to the model. The response from the model is compared against the actual output from a real-system. The performance metric during the validation phase is calculated and compared with the performance metric calculated during the calibration phase. If they are close enough, the model is declared as developed and well-tuned. If the model does not pass the evaluation criteria, it is sent back to the calibration phase for fine tuning. In case of large errors, the model structure and complexity are changed. This may include changing the initial model. The selection process involves choosing the model types with right trade-offs are explained further in Section 3.4.

In case of black-box models, model calibration and identification are used together to obtain the model using part of the training data. The rest of the training data is used for validation purposes.

3.3 Components and Systems

Buildings have complex interconnected systems with several components. The components or the systems need to be modeled for a variety of use cases. The variables of interest depend on the modeling needs and sensor measurements. For example, in the Newton's second law of motion, the force acts as an output if acceleration is measured through sensors, and vice versa. Pumps, fans, valves, and dampers are abundantly utilized in HVAC systems. Pumps are found in chilled water loops, condenser loops, heat recovery systems, boilers, hot water systems, and other refrigerant systems to push the refrigerant. Valves are normally present with pumps in preheating systems, heat recovery, chilled water, hot water, and refrigerant systems. Additional types of valves are reversing, isolation, mixing, bypass, and fill valves. Types of fans are exhaust, supply, return, ventilation, cooling tower, booster, and stand-alone fans. Dampers (if present) are found in conjunction with fans, and thus they are available in the same applications as those in fans. Fire damper is another type of damper in building systems; this damper is associated with fire safety system. The rest of this section describes the possible models of some of the building components and systems.

3.3.1 Fan

1. Third order regression model used in DOE-2 [16] and HVAC toolkit [5] to calculate fan power:

$$\frac{P}{P_{rat}} = C_0 + C_1 \frac{Q}{Q_{rat}} + C_2 \left(\frac{Q}{Q_{rat}} \right)^2 + C_3 \left(\frac{Q}{Q_{rat}} \right)^3 \quad (3.1)$$

where Q is air flow rate, P is fan power, P_{rat} and Q_{rat} are the rated power and rated air flow rate, respectively. This model requires at least four different operating points to find the polynomial coefficients C_0, C_1, C_2, and C_3.

2. In this detailed fan model found in the HVAC toolkit and article [5, 6], the fan performance is characterized in terms of pressure rise across the fan and shaft power. It uses the dimensionless coefficients of flow rate, pressure head, and shaft power. The performance of a fan is represented by a fourth order polynomial with dimensionless coefficients. The coefficients are determined from the manufacturer's data.

3. The fan power (P) is calculated as a function of the fan air flow rate (Q) and total static pressure. The total static pressure is equal to the static pressure set point (P_{set}) plus the remaining duct pressure drop, which is a function of the fan air flow rate. Applying existing system characteristics, the fan energy use is then given as:

$$P = \frac{q}{680000}\left(P_{set} + \frac{2}{1000000}q^2\right). \tag{3.2}$$

Changes in fan power is calculated using the following discretized model [22]:

$$P[k] = P[k-1]\frac{Q[k-1](P_{set}[k-1] + CQ[k-1]^2)}{Q[k](P_{set}[k] + CQ[k]^2)}, \tag{3.3}$$

where $[k]$ represents the variable at time instance k, and C is flow coefficient determined at design conditions.

4. Power as a function of speed is represented by the following [21]:

$$P = cN^3, \tag{3.4}$$

$$P_2 = P_1\left(\frac{N_2}{N_1}\right)^3,$$

where P corresponds to fan power, N represents the fan speed, c is a constant, subscripts 1 and 2 represent two different states of operation.

5. Power as a function of flow rate and static pressure is calculated as [21]:

$$P = \frac{QP_r}{6356}, \tag{3.5}$$

where P, Q, and P_r represent the ideal air horsepower, volumetric air flow rate (cubic feet per minute), and pressure (inches of water gauge) or resistance, respectively.

6. Total air flow rate at an AHU is the sum of total air flow rates at individual zones assuming that there is no infiltration or exfiltration [12].

3.3.2 Chiller

- Enthalpy model: This model calculates the power consumed by a chiller (P_c) using energy-balance equations at the cooling coils inside an AHU. Chiller power is calculated using the following:

$$P_c = c_1 r(H_o - H_s) + c_2(1 - r)Q_t, \qquad (3.6)$$

where H_o and H_s represent outdoor air and supply air enthalpies, respectively, r represents outside air ratio, and Q_t represents the total building load. c_1 and c_2 are calibration constants.

- Energy and entropy model: In this model, the performance of chiller is expressed in entropy and energy balance equations based on the evaporator inlet water temperature, the condenser inlet water temperature, evaporator duty (kW), compressor power (kW), internal entropy, and evaporator duty (kW). Details on the model can be found in the article [29].

- Empirical model: The model is based on three polynomial curves that calculate the capacity and energy input ratio as a function of condenser and evaporator temperatures. It also determines energy input ratio as a function of part-load operation. Model is described in the article [29].

- As describe in article [18], chiller power is calculated using nominal capacity (Q_{cap}), COP, and part-load ratio (PLR) as:

$$P_c = Q_{cap}COP * PLR * T_{adj}, \qquad (3.7)$$

where T_{adj} is an adjustment factor calculated based on condenser and chilled water temperatures. The coefficients of the T_{adj} expression are calculated from the manufacturer's catalog data or common curve-fitting methods through experiments.

3.3.3 Heat Exchanger

Heat exchange rate (Q_{coil}) of cooling coils is expressed as [18]:

$$Q_{coil} = \frac{c_1 m_{sa}^{c_3}(T_{MA} - T_{CHWS})}{1 + c_2(m_{CHW}/m_{SA})^{c_3}}, \qquad (3.8)$$

where T_{MA} is the temperature of mixed air entering the cooling coils, T_{CHWS} is the temperature of chilled water supply, m_{CHW} is the chilled water flow rate through cooling coils, and m_{SA} is the air flow rate of supply air through the cooling coils. In this model, no geometric data of coils is required and three empirical parameters (c_1, c_2, and c_3) need to be identified from manufacturer's catalog data or experimental data.

3.3.4 Pump

Pump power P_{pump} is calculated as [18]:

$$Ppump = \frac{m_{CHW} H_{CHW}}{g_c n_{CHW}}, \tag{3.9}$$

where $n_{CHW} = f(m_{CHW}, H_{CHW})$ equation represents the total efficiency of the pump. Function f can be a polynomial equation, NN model, or any other curve-fitting representation. g_c is a constant and H_{CHW} is the pump head provided by chilled water pump.

3.3.5 Damper

Economizer damper model presented in the article [23] is derived from the detailed damper/valve model in HVAC 2 Toolkit [5] and book [6]:

$$\Delta P = Kq^2. \tag{3.10}$$

In this model, the pressure drop across a damper (ΔP) is related to the mass flow rate (q) by a flow resistance coefficient K. The flow resistance coefficient K varies with the damper position d.

3.3.6 Thermal Models

Thermal models predict the temperature of a substance inside an equipment or space. Below are the methods on thermal models to predict temperature inside a zone given the inputs and ambient conditions:

- Response factor method [19]: Time series data of responses is calculated from a time-series unit pulses based on building properties.

- Conduction transfer functions [19]: Laplace transfer functions relate temperature and heat fluxes in a wall or a layer of the wall to boundary conditions.

- Finite difference [31]: A wall is divided into a finite number of control volumes for which the heat balance equations are solved.

- Lumped capacitance [7, 11]: The electrical analogy is used to model a building element using resistances and capacitances. A capacitance represents the thermal capacitance of (a layer of) a wall.

- Linear parametric model: ARMAX (Autoregressive–moving-average model) is a type of linear parametric model used for determining the coefficients using time-domain data [33].

- Neural network [28]: Neural networks are based on the same functioning principle as the human brain. The relationships between inputs and outputs are determined by linear or nonlinear relationships defined in the neuron layers.

- Genetic algorithm/optimization-based model [20]: Developing a model is formulated as an optimization problem in which the primary goal is to minimize the objective function, which is the difference between the model outputs and true outputs/measurements.

3.3.7 Moisture and Humidity

Moisture/humidity models determine the absolute or relative amount of moisture in air and other fluids. Similar to thermal models, moisture/humidity models also contribute to the overall comfort level. Therefore, they are equally important as thermal models. Uninvited moisture affects the performance and life of an equipment. High moisture content can damage compressors and building materials. Corrosion and mold are also the consequences of high humidity in buildings. In research laboratories, clean rooms, and food industry, the level of moisture content is tightly managed because of narrow operating conditions and strict rules/regulations. Discomfort in the indoor climate because of too high or low humidity can reduce people productivity and cause health concerns.

Most models for humidity fall into grey box model category in which a model is derived from first principles with its parameters estimated/calibrated from experimental data. Many humidity models exist in market today; refer to the studies for additional details [27, 10, 30, 9]. There are a few other model types that use either polynomial equations or lumped parameters, e.g., capacitance-resistance analogy [32]. Some are well tested while others are used only for research purposes. The main dissimilarities between the models are the assumptions, simplicity, and type of available sensor inputs. A couple of model variations in this category are described next.

- In this model [11, Chapter 2] , humidity ratio (W) dynamics are represented by the following equation:

$$\dot{W} = KT \left[n\omega_{H_2O} + m^{SA} \frac{W^{SA} - W}{1 + W^{SA}} \right], \tag{3.11}$$

where n represents the number of people, ω_{H_2O} is the water vapor released by a person, m^{SA} and W^{SA} are the air flow rate and humidity ratio, respectively, of air entering the zone, T corresponds to the zone temperature, and the variable K (which can be treated as a constant) is dependent on the air properties, volume, and pressure. The model is derived from balancing the moisture content of supply air and return air with humidity contribution from occupants.

- Another model is represented by the following equation [34]:

$$\dot{h} = -\alpha h + \beta h_{sat} - k(h - h_{out}) + CQ_{source}, \qquad (3.12)$$

where h represents the zone-specific humidity, α, β are admittance factors, k is the air exchanging rate factor between inside and outside air, Q_{source} is moisture generation rate, and C depends on air density and volume similar to the last model. The $-\alpha h + \beta h_{sat}$ part corresponds to the humidity difference between zone air and interior fabrics during moisture absorption process. Subscripts *sat* and *out* correspond to air saturation and outdoor air, respectively.

3.3.8 Simulation Tools

Numerous software tools are present today to simulate several components of building systems including the indoor environment. A comparison of 20 simulations tools (EnergyPlus^{TM}, HAP, BLAST, TRNSYS, ECOTECT, BSim, DeST, DOE-2.1E, EnerWin, Energy Express, Energy-10, eQUEST, ESP-r, IDA ICE, IES<VE>, HEED, PowerDomus, SUNREL, Tas, and TRACE) in terms of features/capabilities is performed in the review article [8]. Matlab Simscape and Modelica libraries are two other software tools growing in popularity, in addition to the tools mentioned in the article. In selecting a simulation software, open-source vs. closed-source is another attribute that need to be considered along with the other attributes described in Section 3.4. On top of these tools, there are network simulators (NS-1, NS-2, NS-3) [1] and real-time simulators (OPAL-RT) [24] for smart-grid applications with buildings as participants.

As shown above there are several modeling options for every component in buildings. Because of limited scope of this book, only a few options are presented to provide an overview of such components. Although the techniques are represented only for a few components, the same methodology can be applied or extended to other building systems. Modeling details on other components—mixing boxes, pumps, chillers, ducts, humidification, coils, chillers, fan motors, valves, energy modeling methods, cooling towers, and other HVAC components—as can be found in these review articles and studies [3, 4, 30, 36, 25, 2].

3.4 Model Selection

Choosing the right model is one of the most challenging tasks as it requires complete understanding of the problem and the systems involved in the process. Figure 3.3 shows the trade-offs involved in choosing a model as a function of its complexity. On the left side, there are "simple" models with lowest complexity while the "complicated" models with highest complexity are shown on the right side. Trade-off features of simple and complicated models are described below:

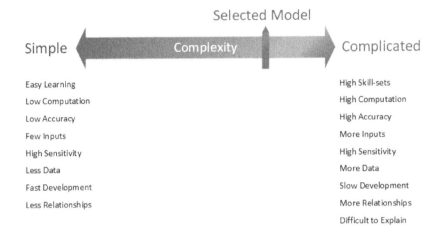

FIGURE 3.3
Trade-off variables as a function of complexity during model selection.

- *Learning and understanding:* Simple models are easy to learn and understand while complicated models need specialized skills and expertise to understand. It means that the cost and effort of learning and understanding is much higher for complicated models. Similarly, simple models are easier to explain than complicated models. In some scenarios, the consumers or intermediate stakeholders are interested in understanding the functioning of systems so that they can interpret results and solve issues by themselves.

- *Computational power:* Complicated models demand much higher computational power than simple models. Using complicated models in a problem formulation imposes constraints on the platform where the model is being deployed. For example, control applications with a complicated model may not be possible on an embedded controller.

- *Accuracy:* Complicated models are expected to be more accurate than simple models. Accuracy of models affects the design of control algorithms in several ways. If a model is accurate, the control time step can be larger or the compensation techniques can be relaxed.

- *Number of inputs:* Complicated models have higher number of inputs than simple models. The number and types of inputs are important aspects for consideration. If a model requires the number of people to predict the IAQ of a zone, an occupancy sensor is needed to count the number of people. Using an occupancy counting sensor could be expensive and practically not feasible in certain applications. Less inputs mean less dependency on the sensors.

- *Sensitivity:* Overall, simple models are usually more sensitive to the errors in inputs as compared to complicated models. Part of the reason is that a complicated model captures all the relationships between many inputs and outputs. However, simple models use surrogate relationships to represent the system behavior. In other words, even though many relationships are needed to describe the system behavior, only a few relationships are present in simple models to maintain simplicity.

- *Data requirements:* If a model needs to be calibrated and validated as in the case of grey box or white box models, complicated models demand a lot more data than simple models. This is because of higher number of relationships and their corresponding coefficients that need be calculated for complicated models.

- *Development speed:* Because of the complicated models' complexity and their additional requirements, their overall development is slower than the development of simple models. Complicated models are more difficult to implement than simple models in many situations.

Complicated models are not necessarily the best models or better than simple models. Selection of a "right" model depends on the application, control algorithm, system architecture, and the product, which further relies on its targeted market, customers, and other stakeholders. Purpose of a model is also very relevant in the selection or elimination process. For example, choosing a white box model from existing simulators (e.g., EnergyPlus) is a good fit for initial development of new control algorithms. However, most of these simulators may not be the best option for real-time embedded controls involving transients and fast dynamics. Sometimes preliminary research is preferred as it is hard to know all the details, trade-offs, and project requirements in advance. In this chapter, the pros and cons of several modeling technologies, methodologies, and fundamentals are explained so that the readers can make the best choice as per their objectives, requirements, and constraints.

> *"I can never satisfy myself until I can make a mechanical model of a thing. If I can make a mechanical model, I can understand it. As long as I cannot make a mechanical model all the way through I cannot understand."*

> William Thomson, *Lord Kelvin*

Key Takeaways: A Few Points to Remember

1. Mathematical models (also called "models") are developed to understand and mimic the behavior of real-systems.

2. Models enable the development and testing of control products before they are deployed on real systems.

3. There are many types of models in multiple dimensions with different attributes such as white vs. grey vs. black box models, linear vs. non-linear models, static vs. dynamic models.

4. Common features and attributes of a model in buildings are transparency, usability, computational power, accuracy, error sensitivity, learning/expertise level, implementation effort, developmental time, deployment speed, operating range, number of inputs, and data requirements.

5. Numerous analytical models and simulation tools exist for individual components and entire systems in buildings. For example, there are more than a dozen models on zone thermal dynamics and there are half a dozen models for fans as shown earlier in the chapter. These models and tools have unique characteristics, trade-off, and selling points.

6. In practice, the is no universal "best-model." Selecting and developing the "best-model" is an iterative process that meets certain performance criteria while staying within the allowed boundaries.

Bibliography

[1] The Network Simulator–ns-2. `https://www.isi.edu/nsnam/ns/`. Accessed: 2020-11-30.

[2] Abdul Afram and Farrokh Janabi-Sharifi. Review of modeling methods for HVAC systems. *Applied Thermal Engineering*, 67(1-2):507–519, 2014.

[3] Seyed Mohammad Attaran, Rubiyah Yusof, and Hazlina Selamat. Short review on HVAC components, mathematical model of HVAC system and

different PID controllers. *International Review of Automatic Control*, 7:263–270, 05 2014.

[4] Rogerio Barbosa and Nathan Mendes. Combined simulation of central HVAC systems with a whole-building hydrothermal model. *Energy and Buildings*, 40:276–288, 12 2008.

[5] MJ Brandemuehl, S Gabel, and I Andersen. A toolkit for secondary HVAC system energy calculation. *American Heating, Refrigeration and Air-conditioning Engineers (ASHRAE)*, 1993.

[6] DR Clark. HVACSIM+ building systems and equipment simulation program: Reference manual. Technical report, United States Department of Commerce, Washington D.C., 1993.

[7] J Crabb, N Murdoch, and J Pennman. A simplified thermal response model. *HVAC&R Research*, 8:13–19, 1987.

[8] Drury Crawley, Jon Hand, Michael Kummert, and Brent Griffith. Contrasting the capabilities of building energy performance simulation programs. *Building and Environment*, 43:661–673, 04 2008.

[9] Steven Emmerich, Andrew Persily, Steven Nabinger, SJ Emmerich, AK Persily, and SJ Nabinger. Modeling moisture in residential buildings with a multizone IAQ program. *Indoor Air*, 9, 07 2002.

[10] Samuel Glass and Anton TenWolde. Review of moisture balance models for residential indoor humidity. In *Proceedings of the 12th Canadian Conference on Building Science and Technology*, volume 1, pages 231–245, 05 2009.

[11] Siddharth Goyal. *Modeling and control algorithms to improve energy efficiency in buildings*. PhD thesis, University of Florida, 2013.

[12] Siddharth Goyal, Herbert Ingley, and Prabir Barooah. Occupancy-based zone climate control for energy efficient buildings: Complexity vs. performance. *Applied Energy*, 106:209–221, June 2013.

[13] Soteris A Kalogirou. Artificial neural networks in renewable energy systems applications: a review. *Renewable and sustainable energy reviews*, 5(4):373–401, 2001.

[14] Soteris A Kalogirou, Georgios A Florides, Costas Neocleous, and Christos N Schizas. Estimation of the daily heating and cooling loads using artificial neural networks. 2001.

[15] Donghun Kim, Yeonjin Bae, Sehyun Yun, and James E Braun. A methodology for generating reduced-order models for large-scale buildings using the Krylov subspace method. *Journal of Building Performance Simulation*, 13(4):419–429, 2020.

[16] Lawrence Berkeley National Laboratory. DOE 2 reference manual. Technical report, United States Department of Energy, Washington D.C., 1980.

[17] Kangji Li, Wenping Xue, Chao Xu, and Hongye Su. Optimization of ventilation system operation in office environment using POD model reduction and genetic algorithm. *Energy and Buildings*, 67:34–43, 2013.

[18] Lu Lu, Wenjian Cai, Lihua Xie, Shujiang Li, and Y.C. Soh. HVAC system optimization–in-building section. *Energy and Buildings*, 37:11–22, 01 2005.

[19] G Mitalas and D Stephenson. Room thermal response factors. *ASHRAE Transactions*, 73:1–10, June 1967.

[20] Melanie Mitchell. *An Introduction to Genetic Algorithms*. MIT Press, Cambridge, MA, USA, 1998.

[21] B. Mleziva. Fan selection and energy savings. *HPAC Engineering*, 82:40–45, 08 2010.

[22] Nabil Nassif, Stanislaw Kajl, and Robert Sabourin. Optimization of HVAC Control System Strategy Using Two-Objective Genetic Algorithm. *HVAC&R Research*, 11(3):459–486, 2005.

[23] Nabil Nassif and Samir Moujaes. A new operating strategy for economizer dampers of VAV system. *Energy and Buildings*, 40:289–299, 12 2008.

[24] OPAL-RT Technologies, Inc. OPAL-RT Real Time Simulator. `https://www.opal-rt.com/`. Accessed: 2020-06-02.

[25] Ahmad Parvaresh, Seyed Mohammad Ali Mohammadi, and Ali Parvaresh. A new mathematical dynamic model for HVAC system components based on Matlab/Simulink. *International Journal of Innovative Technology and Exploring Engineering*, 1(2):1–6, 2012.

[26] Siddharth Goyal Radhakant Padhi, Narayan P. Rao and Abha Tripathi. A method for model-reduction of nonlinear building thermal dynamics. *Automatic Control in Aerospace*, 3, May 2010.

[27] Francesco Scotton. Modeling and identification for HVAC systems. Master's thesis, University of Padua, Padua, Italy, 2012.

[28] I. Seginer, T. Boulard, and B.J. Bailey. Neural network models of the greenhouse climate. *Journal of Agricultural Engineering Research*, 59:203–216, 11 1994.

[29] P. Sreedharan and Phil Haves. Comparison of chiller models for use in model-based fault detection. In *International Conference for Enhanced Building Operations*, pages 1–10, July 2001.

[30] C.P. Underwood and F.W.H. Yik. *Heat Transfer in Building Elements.* John Wiley & Sons, Ltd, 2008.

[31] U.S. Energy Information Administration (EIA). *Energy simulation in building design.* Adam Hilger, Bristol, 1985.

[32] Jason Woods, Jon Winkler, and Dane Christensen. Moisture modeling: Effective moisture penetration depth versus effective capacitance. In *Thermal Performance of the Exterior Envelopes of Whole Buildings XII International Conference,* pages 1–13, 2013.

[33] Siyu Wu and Jian-Qiao Sun. A physics-based linear parametric model of room temperature in office buildings. *Building and Environment,* 50:1–9, 2012.

[34] Jingyang Xu and Daniel Nikovski. A humidity integrated building thermal model. In *American Control Conference,* pages 1492–1499, 07 2016.

[35] In-Ho Yang, Myoung-Souk Yeo, and Kwang-Woo Kim. Application of artificial neural network to predict the optimal start time for heating system in building. *Energy Conversion and Management,* 44(17):2791–2809, 2003.

[36] Xin Zhou, Tianzhen Hong, and Da Yan. Comparison of building energy modeling programs: HVAC systems. Technical report, Lawrence Berkley National Laboratory (LBNL), 01 2013.

4

Control

Control is crucial and a necessary part of any working system. Some systems have high involvement of controls while other systems require minimal level of controls. A system without control is similar to "all talk and no action." Controller is like the brain of a body or CPU (Central Processing Unit) of a computer that gathers all the necessary information, processes the information, and takes the "best" action to meet its goals or objectives. Figure 4.1 shows the block diagram of a closed-loop control system.

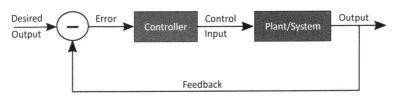

FIGURE 4.1
Block diagram of a closed-loop control system.

In the diagram, the actual output (output) from the system or a plant is fed back to the controller. The actual output is measured from sensors and compared against the desired output. The desired output is normally specified by a user. Difference between the desired output and actual output is called error, which is fed to the controller. The controller processes the error—calculated from the sensors and users—to decide the next control action or control input. Exogenous inputs are non-measurable variables that affect both the plant and controller, e.g., noise, disturbance, and unknown measurements; exogenous inputs are not shown in this case.

Simple example: If the HVAC system at a residence is considered as a plant, the output is the room temperature measured from the temperature sensor embedded in the thermostat. Setpoint provided by the user via the thermostat is regarded as the desired output. Turning on the HVAC or turning off the HVAC equipment is deemed as the control action or control input.

Closed-loop vs. Open-loop Control System: The example in Figure 4.1 shows a closed-loop control system because the response of a system is fed back to the controller in form of error, i.e., deviation from the desired output. If there is no feedback to the controller, the system is called an open-loop control system. Although the value of feedback in any control system is undisputed, open-loop controllers are also prevalent in building systems, mostly at the supervisory level. Schedule-based controls is a great use-case of open-loop controller. In this case, the HVAC or lighting systems are turned on and off at predefined times based on the work schedule, e.g., turning on the HVAC system at 8:00 am in the morning and turning it off at 6:00 pm regardless of the indoor environmental conditions.

4.1   Control Product Types

There are a variety of control products offered by several control manufacturers with different business models to serve diversified customer needs. In this book and this chapter, control product encompasses all the following categories of control products:

1. Control platform is a bare minimum physical hardware (e.g., controller with no algorithm) capable of interacting with a building system and running a control algorithm. Interactions with the building system can be direct or indirect through other hardware. Plug-in computers or Arduino-based micro-controllers are a couple of examples in this area. Specialized, custom servers with/without data acquisitions cards and additional input/output interfaces also fall in this category.

2. Control algorithms and software packages are the tools that can run on application specific controllers, supervisory controllers, or other hardware (e.g., servers and computers) that interact with the controllers or the plant. These packages may include firmware or software updates (e.g., security patches) to run the control algorithms in safe and secure fashion.

3. Physical hardware controllers embedded into the equipment directly controlling the equipment. Control algorithms are being executed in real-time as part of the embedded controllers. Typically, embedded controllers support minimal interface/data-exchange with the external world. There is minimal flexibility to access or change the software or algorithms running on embedded controllers. Such controllers are part of an equipment. It means that the embedded controllers are also replaced when the equipment is replaced.

4. Full-fledged controllers offer complete functioning of control algorithms including the hardware piece as well. The control algorithms can be configurable or fully-programmable at the application level. These controllers are relatively modular in nature and have higher computational power. We can think of them as a specific combination of Items 1 and 2, which has all the bells and whistles to run and operate the equipment. Application specific controllers with loaded VAV control algorithm and standard interface is an example of the controller.

5. Services related to controls, but may not be necessarily part of the main control stream. FDD [26] is a service to monitor the control system performance and report to the building owner or operator in case of current or potential problems. Control consulting, performance contracting [44, 10], retuning the system controls [46], and the services to save energy consumption are also regarded in this category. Services can be local, remote, or through the cloud. Some of the services may be closely related to the control algorithm or software packages.

6. Accessories, special projects, and other products are considered as a separate category that facilitate toward efficient working of the control systems in a building. Retrofitting is a special project, which combines services and control product replacements. Network devices (e.g., gateways, protocol converter, broadcast management device, router, ring manager) to handle the BAS and control communication protocols are a few other cases although they are not directly or indirectly controlling the equipment.

Figure 4.2 summarizes the category of control products.

This chapter provides a detailed overview of control approaches and techniques (in general and used in building systems) along with the selection of right control techniques. The chapter starts with Section 4.2, which explains the value and benefits of controls in buildings. Sections 4.3 and 4.4 briefly explains the approaches and strategies, respectively, used in controls. The chapter ends with Section 4.5 that summarizes the chapter and provides a framework to select the "best" controller.

4.2   Value of Control in Buildings

In building systems, control(s) is of high importance because of the following main reasons:

1. High Involvement: Control is present in almost every part of building systems including lighting systems, power systems, HVAC

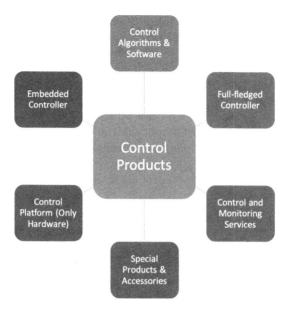

FIGURE 4.2
Types of control products.

systems, refrigeration systems, fire suppression, hot water systems, fire and safety systems. However, the level of controls involvement varies based on the type and purpose of the system. Fire suppression control may be pneumatic with minimal data exchange to an external system (or the cloud) while an HVAC system sensors, actuators, and other components of the system may transfer every data point to the cloud. Chapter 1 has already shown the impact of buildings on everyone's daily life. To improve the overall functioning of the entire building or its individual component/system, it is obvious that upgradation in controls is likely to be a major part of those improvements.

2. Substantial Impact: By improving the controls, one can reap many benefits in the buildings. Controls are capable of reducing overall energy consumption, reducing carbon emissions, changing the internal environment of the buildings by improving the health and productivity of users. Numerous studies and research articles have shown that advancement in controls yield 30–40% energy reduction in both commercial and residential buildings [20, 21, 18, 35, 48, 38]. Moreover, advanced controls can also improve productivity and heath of occupants [16, 40].

3. Low Investment: There are many other ways to reap the above-mentioned benefits. Control, however, in many cases, turns

out to be the most cost-effective way to achieve those goals because of its low investment requirements. For instance, energy consumption in buildings can be reduced by replacing the older HVAC equipment with high-efficient HVAC equipment, which will require huge investment because of high purchasing cost, high labor cost in removing and discarding the old/bulky equipment, and high-installation cost, e.g., new piping or electrical connections. On the contrary, upgrading controls could be as simple as updating software on the controllers or replacing the controllers. In addition to the capital cost, the average time needed to replace the controller firmware, software, or control algorithm is also much shorter than replacing an equipment.

4. Faster Cycle: Control offers faster development and turn-around cycle. It means that if there is a problem with the control algorithm, the update can be quickly developed and implemented on the system (sometimes over the cloud) as compared to fixing a sensor by sending a technician or recalling a large HVAC equipment. Furthermore, the upgradation of controls is likely to be more convenient, easier, and faster than the equipment upgradation. It also means that a few portions of controls are able to accommodate the technological changes and incorporate those changes into the products.

In a nutshell, building controls have societal, environmental, and economic impacts with benefits being offered from multiple perspectives. Where there are so many benefits associated with control, it is expected that there are some challenges with making the advanced or improved controls a reality in the buildings. If there were no challenges, the advanced controls would have already been put in place. Uniqueness of buildings, applicability of controls, cost considerations, diversified stakeholders with different requirements are a few challenges that are explained in Section 8.6 and other parts of this book.

4.3   Control Approach

Control approach needs to be decided before selecting the exact logic and the exact control strategy. Control approach is a crucial part of the control system structure and its architecture. This section discusses a few types of control approaches—distributed, centralized, decentralized, hierarchical—which can be used while implementing them on a system. All these control approaches are used in building systems although some approaches are more prevalent than the others. Sometimes a hybrid approach or combination of control approaches is also deployed on the systems.

4.3.1 Centralized

In centralized control, the controller receives sensor measurements and exogenous inputs at one central location. A centralized controller processes all the information to calculate the control inputs for the entire system. In this system, the entire decision power is owned by a single controller [42, Chapter 6]. Figure 4.1 shows an example of a centralized control system in which all the control functions are performed at one central location. Another perspective of a centralized controller is shown in Figure 4.3. Figure 4.3 shows that there is only one controller that is interacting with the plant to process every information and controlling the entire system.

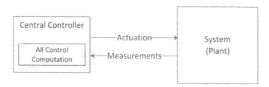

FIGURE 4.3
Simplistic diagram of a system with single centralized controller; exogenous inputs are not shown here.

Centralized control system requires sufficient bandwidth to transfer data from multiple sources such as sensors, actuators, and user inputs. Depending on the control logic, centralized controller may also need high computational power. Furthermore, centralized controller may exhibit high failure rate because of monolithic and centralized location of the control functions. If one function or part of the system breaks, the entire system can be severely affected. Implementation and maintenance in centralized control system are relatively easier than other control approaches. As compared to other control approaches, central controllers possess much higher knowledge of the entire system, and thus can optimize the entire system performance better to reach a certain goal.

4.3.2 Decentralized

Any controller that is not centralized is deemed as a decentralized controller. Figure 4.4 shows a simple schematic of a control system with a couple of decentralized controllers for the entire system. As shown in the figure, the decentralized controllers are allowed to access only local information from the plant to control only a certain part of the system. Decentralized control systems can be thought of a system of subsystems with centralized controllers assigned to each subsystem; basically, there are multiple independent decision makers. Although decentralized controllers don't necessarily share information among each other, it may be necessary that the individual decentralized controllers work properly to ensure the proper functioning of the overall system. This depends on the strength of correlation or interconnections between the process variables that are controlled by individual decentralized controllers.

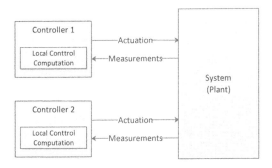

FIGURE 4.4
Simplistic diagram of a system with multiple decentralized controllers; exogenous inputs are not shown here.

In an AHU-VAV HVAC system, there are multiple decentralized controllers that control different components of the system, even within in the same equipment. For instance, there are two independent PI (proportional-integral) control loops in a VAV controller, which decides the damper position and the temperature of air entering the room. Similarly, there are multiple PI control loops in the AHU controller. Although all the control loops are part of the entire (a single) control system, they are decoupled as they don't share information among each other. Decentralized controller is preferred when (1) the communication bandwidth between the plant and the controller is narrow, (2) the physical controller has low computational power, or (3) there are no or minimal needs for the subsystem to interact with the larger system. Decentralized control also helps in breaking the large system into small, manageable chunks of subsystems. However, the controller performance may not be optimal as it lacks the global information to make the best decisions for the overall system.

4.3.3   Distributed

Distributed control system is often confused and interchangeably used with decentralized control system. However, there are subtle differences between decentralized and distributed control systems. In distributed controls, the controllers are allowed to share information among each other, including their internal states and the future decisions. Figure 4.5 shows a simple schematic of a distributed control system.

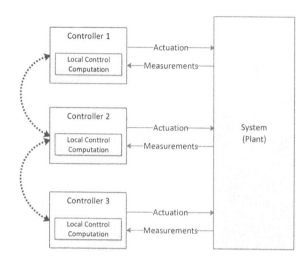

FIGURE 4.5
Simplistic diagram of a system with multiple distributed controllers; a dotted red line shows communication between the controllers.

Similar to decentralized control, distributed controllers assigned to their respective subsystems are making decisions to manage the corresponding portion of the plant. However, they are coordinating among themselves to make better decisions for the entire system. Sometimes, sensor data from other subsystems is also exchanged between the controllers. Furthermore, the computational power is distributed across multiple systems to achieve scalability and faster response time, especially for large-scale complex systems. More the information is shared between the controllers and the plant, the higher is the system orientation toward the centralized control system. Typically, information sharing among the controllers is restricted to their "immediate neighbors" as shown in Figure 4.6. Finding immediate neighbors or the criteria to determine immediate neighbors can be difficult by itself. In buildings, a few options to decide the neighbors are HVAC zone, lighting zone, fire zone, security zone, and physical structure of the building.

In addition to the aforementioned characteristics, distributed control enjoys all the benefits of decentralized control except that development of distributed control not only requires expert knowledge but also could be time consuming. Due to the sharing of information between the controllers, the overall decisions tend to achieve the global minima better

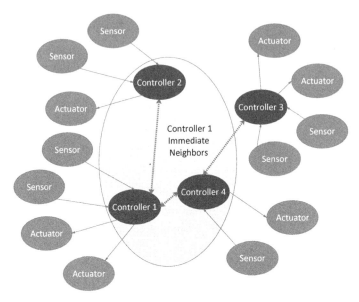

FIGURE 4.6
Mapping diagram indicating the immediate neighbors of Controller 1; dotted red line shows the communication between the controllers.

than the decentralized controllers alone. Distributed control lies in-between decentralized and centralized controls. School of fish and flock of birds are a couple of real-world, natural applications where distributed control can be observed [43, 30, 28]. Distributed control is not much spotted in buildings, but it is an active area of research. Distributed control is expected to take off in upcoming years when the control and applications requirements cannot be easily met with the existing control systems, e.g., buildings are autonomously participating in demand response services as integral part of the large-scale, complex electric grid system.

4.3.4   Hierarchical

In a hierarchical control system, there are many controllers at multiple levels making decisions at different time scales, but ultimately affecting the operation of same subsystem and the plant [33, 12]. The controllers at different levels follow a parent-child relationship. Hierarchical control is also referred to as "layered control" because the control system possesses a layered architecture. These controllers have varying constraints, requirements, and decision power. Figure 4.7 illustrates a generic n-level hierarchical control and an equivalent simplified view of 4-level hierarchical control system found in an AHU-VAV type HVAC system. Details on the AHU-VAV architecture is provided in Section 1.4.

4.3.4.1 Supervisory Control (SC)

It is explained earlier in Chapter 1 that supervisory controller (SC) is present in buildings as part of hierarchical control system. As the name suggests, an SC supervises and coordinates the working of ASCs and embedded controllers. Embedded and application specific controllers equate to local controllers as their responsibilities coincide with each other, i.e., they control the immediate physical device connected to them regardless of the behavior of larger system. SC ensures that local controllers are working together instead of competing with each other. Sometimes, SCs are used to transfer information to local controllers to optimize the overall system. This may include transfer of information on building schedule, setpoints, weather conditions, and other controllers. It is reasonable to think of SCs as complementary controllers used for additional services that are not necessarily needed for basic operations of the system. Instead, SCs act as enablers to provide additional user features and improve the overall functionality of the system. As mentioned earlier, SCs are present in many parts of buildings. Figure 4.8 shows the block diagram of an HVAC control system with emphasis on supervisory controller.

The figure shows a cascade of controllers that manipulate the plant, which is represented by the building dynamics block. Building dynamics are affected by a change in the equipment dynamics, which are controlled by an equipment controller; the equipment controller is the same as application specific controller. A single equipment controller or multiple equipment controllers are further controlled or coordinated by a system-level controller. Each controller has a predefined set of parameters and constraints that need to be satisfied or utilized in the control algorithms. In general, supervisory controllers can (1) decide the fixed parameters used in the controllers, (2) decide the constraints, (3) act as information exchangers between the controllers, and (4) pass global measurements or information to the controllers. In relation to the hierarchical view shown in Figure 4.7, equipment controller is in the 1st layer closest to the actual system while the HVAC system controller is in the 2nd layer. Based on the network configuration and capabilities of equipment controllers, both equipment controllers and HVAC system controllers can be considered as ASCs. As an example of AHU-VAV system, VAV-box and AHU are the equipment with their corresponding equipment controllers, i.e., AHU controller and VAV-box controller. Since there are many VAV-boxes connected to one AHU, the AHU controller acts as an HVAC system controller.

A combination of multiple control approaches is also possible in the building systems. Decentralized and hierarchical controls are two approaches often found in building systems. With improvement in networking, sensing, cloud, and other software technologies, the control system architecture in the

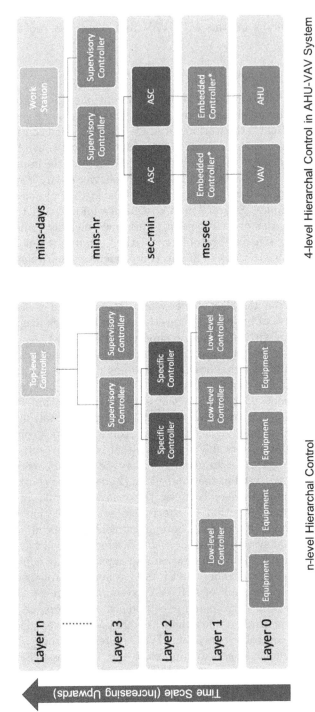

4-level Hierarchal Control in AHU-VAV System

n-level Hierarchal Control

FIGURE 4.7

General schematic of n-level hierarchical control and an equivalent simplified view of 4-level hierarchical control in an AHU-VAV HVAC system; * denotes that the controller may not be needed or may not be available.

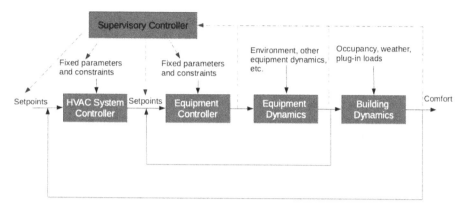

FIGURE 4.8
Block diagram of an HVAC control system with comfort as a primary output.

next-generation set of buildings is most susceptible to change in the nearby future.

4.4 Control Algorithms/Strategies

Once the controller receives all the information, its most critical task is to make decisions for the plant; the decisions are called control inputs or control actions. The logic or the algorithm to drive the decisions are discussed in this section. Control algorithms and control strategies are used interchangeably in this book. Each control algorithm itself is a diverse, large topic involving decades of experience, research, and applications. In this section, only certain aspects that are relevant to building systems are presented.

4.4.1 Bang-bang Control

Bang-bang control is one of the simplest feedback controllers. Bang-bang controllers have two states, i.e., the controller makes decisions between two variables such as on or off, true or false, 1 or 0. The control input u at current time instance takes the following form:

$$u = \left\{ \begin{array}{c} 1, \text{if the condition is true} \\ 0, \text{otherwise} \end{array} \right\}, \qquad (4.1)$$

where 1 and 0 are two possible decisions. The decisions can be either deduced as a function of a time or calculated based on certain conditions. The biggest advantage of a bang-bang controller is its simplicity. It is not only simple to design, maintain, operate, and debug but also simple to explain to others. However, bang-bang controller may show high oscillations, overshoots, hysteresis, and control energy. It may also degrade the equipment performance because of frequent switching between the operating states. Therefore, the bang-bang controller lags behind in terms of overall control performance.

Bang-bang controller is very common in building systems, particularly for simple applications. In lighting systems, a set of lighting fixtures in a specific area is turned on when an occupant or movement is detected in the area, e.g., supermarkets. After a certain time of inactivity, the lights are turned off. In hotel rooms, lights are turned on when the key is pushed into the key slot. These types of simple bang-bang control algorithms are widely implemented in many different types of buildings.

Bang-bang control is also found in many HVAC systems, primarily when the control options are limited to only one control input with two states, on and off, e.g., roof-top units. Figure 4.9 shows the working of bang-bang controller for a single-stage/speed air-conditioning system in homes. This system is equivalent to a heat pump system but allows only cooling, i.e., there is no reversing valve. Heating is provided by another source such as furnace and baseboard heater. In summer, the system is in cooling mode with a temperature setpoint T_{sp} provided by the occupant. When the room temperature passes above the setpoint by a certain value (represented by deadband), the controller turns on the air-conditioning system until the room temperature falls below its lower limit. When the room temperature falls below the lower limit, the air-conditioning system is turned off until the room temperature reaches its upper limit.

4.4.2   Rule/Event-based Control

Rule-based control is a broad classification of control algorithms [18, 36]. A rule-based controller determines the control actions based on certain rules or "if-else" conditions rather than continuous analytical mathematical equations. The control action can be expressed in the following form:

$$u = \left\{ \; A_i, \text{if } C_i \text{ is true and } C_j \text{ is false} \; \right\}, \quad \begin{array}{l} \text{where i } =1,2,\dots,\text{n}, \\ \text{and j } =1,2,\dots,\text{i--1}. \end{array} \quad (4.2)$$

C_i denotes the ith condition and A_i denotes the action associated with the ith condition. The above equations simply mean that there are series of sequential conditions or rules (Condition 1, Condition 2, ...) with their associated actions (Action 1, Action 2, ...). If the current condition is met and the former conditions are not met, an action corresponding to the current action is taken. Sometimes rule-based control is also referred to as an event-based control although event-based control is a subset of rule-based control. In event-based controller, a control action is taken when an event is

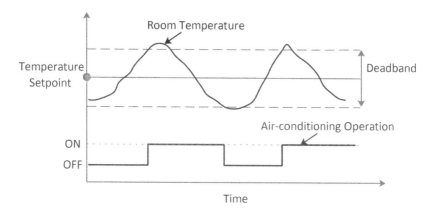

FIGURE 4.9
Working of a bang-bang controller operating a single-stage air conditioning system in residential buildings.

triggered. Rule-based controller or even-triggered controllers are very common in building systems. At supervisory control-level, lights can be turned off during weekends unless anyone enters the building during that time. Similarly, the room temperature setpoint can be changed during unoccupied time period detected/triggered by an occupancy sensor.

4.4.3 Finite State Machine (FSM)

In a FSM algorithm, the system or subsystem operation is decomposed into a finite number of states and the controller decides the next state (or the action corresponding to the next state) and the output (optional) based on the current state and optional measurements/inputs [11, 6]. Consider that there are n states with i^{th} state represented by S_i. An input mapping matrix $(I \in n \times n)$ represents the value of inputs that are used to transition between the states. If the system is in S_i state and input I_{ij} is applied, the system will transition to the next state S_j. The system is allowed to be only in one state at a time. Instead of inputs, a set of conditions can also be used to determine the next state by using a condition mapping matrix (C_{ij}). Figure 4.10 illustrates the working of a simple, three-state $(n = 3)$ FSM logic.

Figure 4.10 shows that there are three states S_1, S_2, and S_3 with their associated inputs and condition mapping matrices. If the system is in S_1 state, applying an input I_{13} will transition to the next state S_3. The same transition happens when the condition C_{13} is met. Similarly, applying input I_{12} will

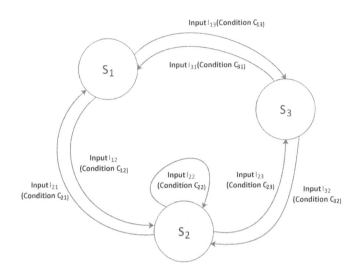

FIGURE 4.10
Schematic of a simplified 3-state finite state machine (FSM).

switch the current state from S_1 to S_2. Once the system is in S_2 state, it remains in the current state when input I_{22} is applied. Outputs are not shown in the figure for ease of convenience. Outputs are variables that can take finite set of values based on inputs and states. The outputs can either be a part of the condition matrix and drive the next state or be supplied as inputs to another control process. Historical data from the states and inputs can also be used while determining the output or the next state of the system. Feedback from states or inputs can also be incorporated into the FSM problem formulation.

Similar to a rule-based controller, the main advantage of FSM is its simplicity. FSM is convenient to use, simple/fast to develop, easy to be understood, and straight-forward to explain to others. Graphical languages and software tools make the algorithms easy to display, comprehend, and use. Another advantage of an FSM controller is that it is easy and fast to verify and test the functionality of FSM algorithms. However, an FSM algorithm can become highly complex and unmanageable if there are too many states. The stability and control performance may not be guaranteed, but the computational requirements are typically very low for such control algorithms.

FSM is prevalent in many parts of building control system [7, 37, 24] because of its above-mentioned advantages. Patent [24] shows the switching of heating, cooling, and satisfaction modes in a VAV box controller according to the conditions utilizing several inputs such as room temperature, cooling and heating setpoints, lockout and saturation conditions.

Figure 4.11 shows a simple FSM diagram used in a VAV-box controller with three states/modes, heating, cooling, and deadband. If the room temperature (T_r) is higher than its cooling temperature setpoint (T_{CSP}) by more than an allowed value (D), the controller state switches to the cooling mode. The controller transitions from cooling to deadband mode if the room temperature is lower than the cooling temperature setpoint by a certain allowed value. The controller switches from deadband to heating mode if the room temperature is lower than its cooling setpoint by a certain allowed value D. The controller is not allowed to switch directly from heating to cooling or vice-versa. Instead, the transition occurs through the deadband mode. The FSM in a VAV controller can involve many more states and the transitions based on the control logic and problem complexity. Furthermore, the transitions may be dependent on historical data and time conditions. For example, the controller waits for a few minutes before it switches the state; this is done to reduce the number of oscillations and bring somewhat stable behavior in the system.

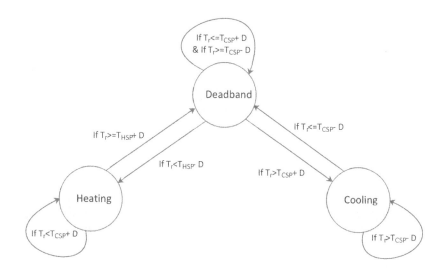

FIGURE 4.11
FSM diagram of a simple VAV box controller of a simplified 3-state finite state machine (FSM).

4.4.4 PID (Proportional-integral-derivative) Control

PID controller is one of the most common and practical control algorithms used in many industries, not just buildings. As the name indicates, PID controller has three main components [3]:

1. Proportional: This component $(K_P e(t))$ corresponds to the input resulting from the deviation of process variable from its reference/setpoint. Process variable is the system output that is being manipulated by the controller actions. Farther is the process variable from its setpoint, the higher is the input needed to bring the system closer to the setpoint. The entire proportional term involves a constant (K_P) and the error $(e(t))$ term, which is the deviation from its setpoint, i.e., e = process variable − setpoint. This term is considered as the "current" part because it uses the data at the current time.

2. Integral: Integral component $\left(K_I \int_0^t e(t)dt\right)$ incorporates the historical error responses into the controller. Although integral term helps in reducing the steady-state error, it may cause sluggish response from the controller in achieving its setpoint. This part is also considered as the "memory or historical" part of the algorithm.

3. Derivative: This term $\left(K_D \frac{de}{dt}\right)$ uses the derivative of error in calculating the control actions. This term accounts for the trends in the input change over time, and thus interpreted as the "future" term. As an example, if the slope of error term is high, the error is increasing at a faster rate. Therefore, the controller should increase its action to compensate for the increasing error. Derivative component results in faster response time, but may cause higher overshoot.

Combining the three components, the control action determined by a PID controller in a continuous form is expressed as [15]:

$$u = K_P e + K_I \int^t edt + K_D \frac{de}{dt}, \qquad (4.3)$$

where K_P, K_I, and K_D are called proportional, integral, and derivative gains, respectively. PID controller is simple to implement because of available tools and techniques that have been developed and tested over many decades. In building systems for at least last couple of decades, PID logic had been used for many processes that require tracking or regulating a process variable. Therefore, the building industry and its stakeholders are well-aware of the high-level concept and the terminology used in PID controllers. Many sensors used in building are prone to errors because of their low resolution, low accuracy, and high sensitivity to noise/disturbances. When the sensors measuring the process variables exhibit high frequency noise, they cause the derivative term in the PID controller to exhibit higher oscillations. These high frequency oscillations cause unnecessary fluctuations in control actions and outputs. In practice, therefore, the derivative term is usually ignored in PID controllers unless there are very hard requirements on the achieving a small response time. Low-pass filters can be implemented in such scenarios [8]. A practical PI controller present in building systems in discrete form is written in the following form:

$$u(k) = K_P e(k) + K_I \sum_{j=k-N}^{k} e(j), \qquad (4.4)$$

where $u(k)$ represents the control action at time instance k. Constant N corresponds to the number of time instances used for integration purposes because it may not always be practical to integrate the error term indefinitely. Despite its natural advantages, PID logic has inherent disadvantages while handling not only the systems with high non-linear behaviors but also simple constraints for any systems. Therefore, PID controllers are used in conjunction with other rule-based or FSM control algorithms. Tuning the controller gains is one of the biggest challenges for PID controllers. Several methods (e.g., Ziegler-Nichols method, frequency-based and step-response) [13, 14] are available to determine the gains. If the system/equipment dynamics for a certain PID loop does not vary much from one building to another, the gains for the loop can be calculated offline and pre-loaded into the controller. This process is quite common for the PID controllers in buildings. Cascaded PID loops are also found mostly in HVAC systems. In cascaded PID control loops, the control action calculated from one controller acts as a reference/setpoint for the other controller. Figure 4.12 shows the schematic of single-maximum strategy [21], which is a typical control algorithm present in VAV-box controllers.

Single-maximum is a combination of PID and FSM algorithms. There are two independent PID loops (heating loop and cooling loop) and three system modes (heating, cooling, and deadband). Figure 4.11 in Section 4.4.3 shows the FSM policy that enables the switching of the modes. The process variable is the room temperature during all the modes. Reference or the setpoint during cooling mode is the cooling temperature set point. In cooling mode, the control input is the supply air flow rate and only cooling PID loop is active. It means that heating valve is turned off and only cold air supplied from its AHU is directly discharged into the zone. Amount of air supplied to the zone is calculated using the PID logic. The SA flow rate calculated in this loop is fed as a reference to another PID loop, which calculates the damper position as the damper in the VAV box changes the SA flow rate. This is an example of cascaded PID loops.

In heating mode, heating PID loop is activated while keeping the supply air flow rate at its minimum value. In the heating control loop, the control input and the reference are SA temperature and heating temperature setpoint, respectively, i.e., the controller changes the SA temperature to maintain the heating temperature setpoint. In some cases, an average of heating and cooling temperature setpoints are used as a reference for both PID loops.

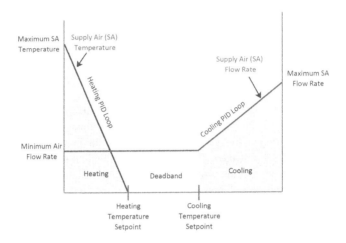

FIGURE 4.12
Single-maximum control algorithm in VAV boxes.

In the deadband mode, the SA flow rate and SA temperature are kept at their minimum allowed values, i.e., heating is turned off. Minimum air is supplied to satisfy ventilation requirements; minimum air flow rate is normally 30–40% of the designed maximum air flow rate. SA temperature calculated from the heating PID loop is fed as a reference to another PID loop, which calculates the heating valve position. The heating valve position modulates the SA temperature. An advanced version of single-maximum is called dual-maximum [21], which is shown in Figure 4.13.

In contrast to the single-maximum, dual-maximum has four operating modes (reheating, heating, cooling, and deadband) and the minimum air flow rate is much lower (10–15% of designed maximum). Reheating mode occurs when the zone temperature drops below the reheating temperature setpoint by a certain value. In reheating mode, the reheating PID loop is activated and SA temperature is kept at the maximum value, i.e., the heating valve is fully open. In the reheating loop, the process variable (room temperature) is the same as in other modes, and the control input is the SA flow rate. It means that the supply air flow rate is modulated to track reheating temperature setpoint. In AHU, chiller, and other building systems that require tracking or regulating a setpoint, PID or PID-like algorithms are used. A few applications of PID loops are:

• Conditioned air temperature loops: To maintain the temperature setpoint of conditioned air leaving the cooling coils and heating coils inside an AHU.

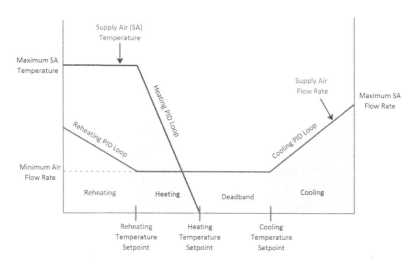

FIGURE 4.13
Dual-maximum control algorithm in VAV boxes.

Control inputs for cooling and heating coil loops are cooling valve positions and heating valve positions, respectively.

- Differential pressure loops: To maintain the differential pressure setpoints in chilled water and hot water loops. Primary/secondary chiller water pumps and hot water pump speeds are controlled to maintain the setpoints. Bypass valve positions can also be controlled to affect the differential pressure in the loops. Similar algorithms are applicable to the condenser loops in cooling towers.

- Air flow rate loops: To maintain a certain air flow rate supplied by outside and supply fans. The control inputs can be the fan speed or the damper position.

4.4.5 Model Predictive Control (MPC)

MPC is a control algorithm in which the control inputs at current time are obtained by solving an optimization problem over a certain period of time in the future (called "horizon"). The control inputs at the current time are implemented on a real-system, and the optimization problem is solved again at the next time instance considering the updated system response after the last input had been implemented. MPC is also referred as RHC (Receding Horizon Control) because the horizon keeps on moving every time a control

input is implemented. As the name indicates, MPC needs the plant model
to obtain predictions, which are used further to calculate the control inputs.
MPC calculates the best input $(u(t^*))$ trajectory that satisfies all the necessary
constraints given the model dynamics. Figure 4.14 illustrates the working of
MPC in abstract form. Solid lines show the actual system state or applied
input while dotted or dashed line shows inputs and their associated predicted
states.

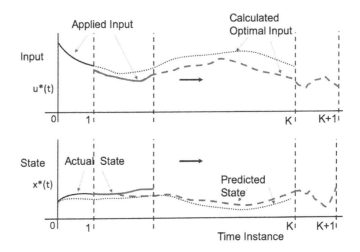

FIGURE 4.14
Graphical workflow of an MPC algorithm.

At the current time (time instance 0), optimal input trajectory $(u(t^*))$
for the time horizon K is calculated. The corresponding system state—which
is the overall state of the system (including its outputs) if the optimal input
trajectory is applied to the system— predicted by the model is also calculated.
At current time, the optimal input trajectory and its corresponding state are
shown by the black dotted lines. The optimal input is applied at the current
time step as shown by the solid black line. The actual state after applying
input is different from the predicted state from the model because of model
uncertainty, noise, and disturbances present in the system.

At the next time step (time instance 1), the time horizon is moved to $K+1$.
The actual system state, instead of previous predictions, is used to calculate
the optimal input trajectory, which is represented by the dashed orange line.
Similar to the previous step, the first input from the trajectory is applied,
and the actual state of the system for the orange input is noted. The actual

state, which is shown by the orange solid line, differs from predicted state. The entire process is repeated. In short, MPC boils down to the following sequential steps:

1. Solve an optimal control problem to obtain the "best" control actions as a function of time.

2. Apply the first set of control inputs.

3. Obtain the new output/state from the system.

4. Start with the new initial condition determined from the new output of the system.

5. Repeat the entire process.

Optimization problem solved in MPC to determine the optimal input trajectory can be formulated as the following:

$$U^* := \quad arg \min_{u_1,...,u_K} \sum_{k=1}^{K} J(x_k, u_k),$$
$$\text{subject to } f(x_1, ..., x_K, u_1, ..., u_K) \leq 0$$
$$\text{and } c(x_1, ..., x_K, u_1, ..., u_K) = 0,$$

$$(4.5)$$

where J is an objective function minimized by the controller. Functions f and c denote inequality and equality constraints, respectively, within the system. These constraints arise from limits on the inputs, model predictions, stability criteria, robustness guarantees, or performance tradeoffs. Basically, the solution to the above problem minimizes the objective function while satisfying a set of pre-defined constraints.

MPC had been gaining traction in buildings, primarily for HVAC systems both in academia and industry [47, 18], over the past one and half decades because of two main factors: (1) its natural ability to minimize an objective function and handle linear/non-linear constraints simultaneously and (2) technical and business advantages during operational phase by improving comfort and lowering utility bills through reduction in energy consumption and management of peak demands. However, the adoption of MPC in the market is increasing at slow rate because of its complexity, high initial cost during installation and setup, and lack of awareness in the stakeholders. From a pure technical perspective, the following are the main hurdles of developing a good MPC solution in buildings:

1. High computational power: In general, MPC requires higher computational power than a PID or a rule-based logic. With increasing number of components and constraints, the computational power needed by the controller increases substantially. This requirement cannot be met by current application-specific controllers

or supervisory controllers. Therefore, the deployment of MPC is restricted to the cloud or workstation with decent memory configuration.

2. Global minimum and convexification: The nature of building dynamics and objective function causes the overall problem to be non-convex in the decision variables. It means that the control solution may not be the global minimum ("best"). In relation to achieving the best solution in a short-period of time, the problem needs to be modified and formulated in the right form. This may need a linear approximation or creating a convex hull of the original problem. The entire approximation/convexification process is a tradeoff between accuracy and approximation because the new modified problem may not be accurate representation of the original problem [18, 17].

3. Mathematical models: As MPC demands the predictions of exogenous inputs and states over the horizon, a model is needed for every such variable. Accuracy of model dictates the overall performance of control solution. For some variables, a generic model based on assumptions can be adopted. However, it is quite challenging to develop one modeling solution that fits the entire problem because of distinct characteristics of every building and their equipment configurations. For example, occupancy measurements at the future time can be assumed same as the current measurements for office-type spaces, but not for spaces with high variations and fluctuations, e.g., casinos or hallways.

4. Fallback mechanism: It is possible that MPC is not able to calculate a solution or produces an infeasible solution within the allowed period of time. In that case, a set of fallback control inputs need to be generated. A good amount of effort is needed to develop a mechanism to calculate the fallback inputs in the allowed time for not only providing a smooth transition between the changing inputs but also generating a solution closer to the optimal solution.

There are several other technical difficulties in realizing a good MPC solution such as solution sensitivity, solution performance in case of sensors unavailability, lack of sensing information, combining objective functions with different dimensions, solution impact on equipment life-cycle, selection of an appropriate time horizon, providing right weights to different parts of objective function. Despite all the technical and other challenges, *MPC is still a very good candidate for achieving advanced controls in buildings provided right technical and business approaches are adopted.*

MPC can be implemented at a building level, entire system/site level (cooling tower, chiller, boiler, AHU, etc.) or a sub-system level (e.g., AHU-VAV). Depending on the problem formulation with different objective function and constraints, there are many variations of MPC algorithms such as economic MPC, robust MPC, non-linear MPC, hierarchical MPC, stochastic MPC, and distributed MPC [31, 27, 34]. Figure 4.15 shows examples of MPC.

In the figure, there are three examples of MPC formulation at different levels. The first two examples are focused on reducing energy consumption while the last example limits the power demand so that the building owner can reduce the peak-demand costs or provide a service to the power grid market. Although the problems are expressed at high-level in simple sentences in the examples, they can be formulated into mathematical optimization problem. For example, zone-based VAV control example can be expressed mathematically as:

$$\min_{T_{SA}(k), \dot{m}_{SA}(k)} \sum_{k=0}^{K} Energy^2(k) \tag{4.6}$$

subject to:

$$\left. \begin{array}{l} T_{low}^{SA} \leq T_{SA}(k) \leq T_{high}^{SA} \\ 0 \leq \dot{m}_{SA}(k) \leq \dot{m}_{high} \end{array} \right\} \text{Actuator Constraints} \tag{4.7}$$

$$T_{low} \leq T_{room}(k) \leq T_{high} \quad \} \text{Comfort Constraints}$$

$$\dot{m}_{SA}(k) \geq Cn_{people} \quad \} \text{IAQ Constraints}$$

$$T(k+1) = f(T(k), u(k)) \quad \} \text{Dynamics (Model) Constraints,}$$

where T_{SA}, $\dot{m}_{SA}$, T_{room}, and n_{people} denote the SA temperature, SA flow rate, zone temperature, and number of people in the zone, respectively. Constant parameters T_{high}^{SA}, T_{low}^{SA}, $\dot{m}_{SA}^{high}$, T_{high}^{SA}, T_{low}, T_{high}, and C are parts of the constraints. Function f identifies the thermal dynamics model of the zone. This is a simple representation of the problem; additional constraints and modeling specifications such as hygro-dynamics and time-varying constraints can also be incorporated to capture the real situation and improve the controller accuracy.

4.4.6 Adaptive Control

Adaptive control is a set of methodologies to consider the unaccounted and changing behavior in the overall system, including its controller. The unaccounted behavior arises because of the following main conditions:

- Measurement noise from sensors due to precision, accuracy, and reliability limitations, e.g., room temperature sensor producing a value of 77.0 °F (25.0 °C) while the true temperature value is 76.76 °F (24.87 °C).

- Actuation disturbances and errors while controlling plants.

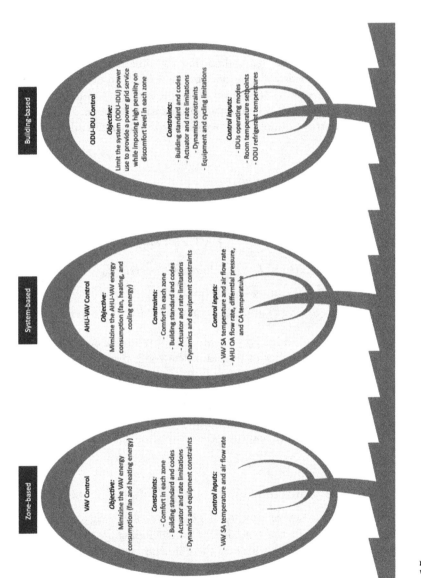

FIGURE 4.15
MPC problem formulation examples at multiple levels.

- Unknown and unmeasured exogenous inputs such as load/occupancy measurements, door status, and window status indicating whether they are open or closed.

- Model mismatches from actual system because of change in operating range.

- Unknown and high-varying model parameters, e.g., seasonal changes.

- Dynamics changes over time, e.g., building materials degradation because of aging and rough environmental conditions.

- Building modifications, e.g., laboratory converted into office.

Adaptive control can coexist with the other control algorithms. The goal of an adaptive controller is to produce a better outcome in such uncertain situations by adjusting itself accordingly. The biggest advantage of adaptive controllers is that they outperform their counterpart, i.e., non-adaptive controllers, because of their ability to handle uncertain and unknown situations. However, it may be necessary to know the type and assumptions involved in those situations. For example, if it is known that the system in linear in nature taking a form $\dot{x} = ax + u$, it is important to know/assume that a is a fixed unknown parameter. An adaptive controller can be designed for such a system without even estimating the parameter a [29]. However, designing such a controller adds a lot more complexity in terms of not only development time and expertise but also testing the control algorithm.

It is also time consuming to design an adaptive controller to tackle all the situations at once. Therefore, the situations that either are happening frequently or have huge impact on the existing controller performance are prioritized. A few strategies used in adaptive control are gain scheduling, adaptive pole placement, Lyapunov-based control, MRAC (model reference adaptive control), and adaptive backstepping [23, 9]. Adaptive controllers are very common in aerospace applications. In building systems, adaptive control is an active area of research [32, 9, 41] although a few simple adaptive control strategies are present in buildings for at least a couple of decades. Two main categories of adaptive controller found in existing buildings are parameter tuning and reference tuning, which are described next.

4.4.6.1 Parameter Tuning

In this strategy, the parameters of existing controllers are updated based on new information or changing environmental conditions. Figure 4.16 shows the block diagram of a control system that changes the parameters of an existing controller. The figure shows that the output and error values from the system are fed into an adaptive law strategy, which updates the parameters of the controller. There may be other external inputs supplied as part of the adaptive law strategy.

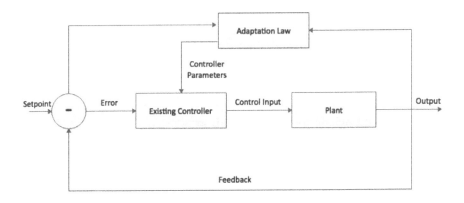

FIGURE 4.16
Block diagram of parameter tuning adaptive control.

There are many forms of adaptive control algorithms to determine the parameters. Since PID is a typical control algorithm used in building systems, one of the most common adaptive strategies is to update the parameters of the PID controller. The parameters are proportional gain, integral gain, derivative gain, integral time, and time step. As the building and equipment dynamics are non-linear, changing the PID parameters may yield a high controller performance in terms of control energy and desired output. For example, SA flow rate varies non-linearly as a function of damper position. If the proportional gain is tuned to a high value (e.g., tuning is performed based on the startup data when the room temperature is far from its setpoint), the damper positions calculated from the controller may show high oscillations of the control action and output.

Usually, the PID parameters are manually determined and set once during commissioning or installation phase. However, building performance changes over time because of degradation in materials and equipment. Moreover, in an entire year, there are several operational and seasonal variations, e.g., summer and winter. Therefore, one set of manually-calculated gains are not sufficient to achieve consistent, optimal performance. PRAC (Pattern Recognition Adaptive Control [25]) algorithm by Johnson Controls is an example of adaptive control, which recognizes patterns in the system behavior and update the PID parameters at regular intervals. Another example is a multi-layer NN-based adaptive control algorithm developed by Siemens [39], which uses setpoint, output, and response time to improve the controller performance.

4.4.6.2   Reference Tuning

Reference tuning method calculates the setpoints which are provided as a reference to the existing controllers. Setpoints can be changed sporadically or frequently to ensure best performance of the system as the plant and controller performances vary over time. Figure 4.17 shows the block diagram of a reference tuning controller. Similar to parameter tuning, adaption law in reference tuning also receives error, output, and possible external inputs such as time, weather, etc. As opposed to parameter tuning, only setpoints are calculated and updated in reference tuning. There are no changes or updates in the controller parameters and the controller structure.

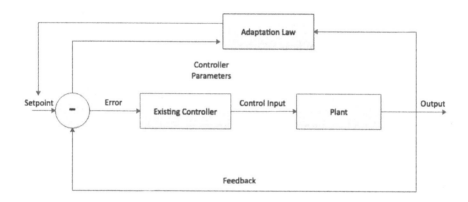

FIGURE 4.17
Block diagram of reference tuning adaptive control.

Adjusting the setpoints or reference signals only results in much higher controller performance because of changes in many factors including environmental trends over the past few years: (1) working schedules, e.g., people flexing their start time and starting to work from home a lot more in recent years, (2) seasonal and daily weather variations, (3) people preferences with age, time, activity level, and environmental factors, (4) equipment best performance and working conditions, and (5) building (infiltration and exfiltration) and insulation levels. It is also common that the setpoint is changed once manually due to user complaint or one special event, but the setpoint is never changed back to its original settings. As a result, many service/software companies have emerged over the past decade to identify potential problems in building systems and provide recommendations along with tuning measures to improve the overall health and performance of buildings. Learning thermostats [2, 1] are great examples of reference tuning control algorithms in residential buildings.

4.4.7 Economic control

Economic control is a group of control strategies that are derived from economics. Economic control uses the concepts of supply, demand, and marginal utilities [4, Chapter 7] to find the market equilibrium or to change the market equilibrium. An example of economic control is to choose the right quantity of items that a person wants to purchase within a given budget. Suppose you have $5 and you have options to purchase apples and bananas. The goal here is to maximize your satisfaction level by purchasing the apples and bananas costing less than $5 in total. Suppose that the cost of an apple and banana are $1.0 and $0.5, respectively. If the satisfaction (not including the price) from every single banana is higher than the satisfaction from every single apple, the obvious solution is that you will purchase only bananas. If the satisfaction from apple is more than twice the satisfaction from banana, the obvious decision is to purchase apples only.

The satisfaction from purchasing a product (an apple or a set of apples) is called total utility. Marginal utility corresponds to additional satisfaction from purchasing or consuming an extra product. In this scenario, the change in utility from purchasing an additional apple is considered as marginal utility. Marginal utility (and marginal utility per cost), derived from the person's satisfaction function as a function of utility functions, can be used to find the right mix of items; refer to article for additional details and example [4, Chapter 7] on this algorithm. Economic control can be applied in centralized, distributed, decentralized, or hierarchical fashions.

Although economic control strategies are not common in buildings, the concept can be adapted to certain building systems. Suppose the HVAC system is turned on around 8:00 am in the morning before the occupants enter the building. AHU has a limit on the maximum amount of air that it can deliver because of physical constraints associated with its supply fan [22]. All the VAV boxes require high amount of air for their corresponding rooms because of high demand as a result of sudden, simultaneous load increase in the morning. Marginal utility functions of each room can be used to maximize the total satisfaction level and find the best possible solution. The satisfaction level can be a function of many factors such as deviation of temperature from its setpoint, type and functions of the room (sales meeting vs. lab testing), and occupant type (internal employee vs. customer).

4.4.8 Miscellaneous

There are numerous control algorithms applied to a variety of systems in different areas and industries [5] ranging from stock market to networking systems to satellite and aircraft systems. Because only few control algorithms are implemented in practice in buildings, advanced control offers an untapped opportunity for both research and business purposes. The following list names few other control algorithms and control methodologies (a set of control algorithms) available in general to the community:

- Nonlinear Control
- Multi-variable Control
- Robust Control
- Stochastic Control
- Optimal Control
- Hybrid Control
- Optimization-based Control
- State-space Control
- Consensus-based Control
- Cooperative Control
- Game Theoretical Control
- Fuzzy Control
- NN Control
- Classical Control

- Lyapunov-based Control
- Observer-based Control
- Sliding Mode Control
- MRAC (Model Reference Adaptive) Control
- Adaptive Backstepping
- Feedback Liberalization
- Approximate Dynamic Programming
- Fault Tolerant Control
- H_2 Control
- H_∞ Control
- Linear Quadratic Regulator (LQR)
- Linear Quadratic Gaussian (LQG)
- Iterative Learning Control
- Extremum Seeking Control

4.5 Optimal Selection of Controls

Optimal selection of controls means choosing the right control architecture, approach, and algorithm to possess the desired system characteristics, behaviors, and features. Because the characteristics/behaviors/features have direct or indirect influence on the product cost, they affect the business decisions in a company. The characteristics along with the tradeoffs between different choices are explained next:

- *Sensing:* Sensing is an absolute need for any feedback control system. Without sensing, any controller will act as an open-loop controller. As the number and types of sensors needed to implement a control algorithm increase, the product cost increases. If a control algorithm requires a measurement that is not present or accessible in the building system, the corresponding sensor needs to be either embedded into the controller or separately installed and wired into the controller; this increases the installation cost as well. Furthermore, in some cases when the sensors cannot be installed because of physical/aesthetic constraints, the number of buildings where the controller can be implemented decreases. Therefore, it is better to design a control system with low sensing requirements. In most buildings, however, only certain/essential sensors are available. Therefore, any advanced sensing is likely to need new sensors that are not currently installed in buildings. Types of sensor requirements also affect the cost. A few questions to consider along those lines are:

- Sensor in equipment, can the sensor be embedded in a controller?
- Location of installed sensor, does the sensor need be installed on the ceiling?
- Installation procedure, is an electrician needed to install the sensor?

- *Modeling:* Mathematical modeling is needed only for model-based controllers. Models correspond to building dynamics, equipment dynamics, or other physical properties of the system. Attributes of model have already been discussed in Chapter 3. Attributes of model directly affect the controller design. Model-based controller usually have high demands as compared to non-model controllers. Higher modeling requirements lead to higher cost and higher time investment into control product.

- *Learning and training:* Simple control products with simple control algorithms are easier to explain to customers and internal teams. Launching a control product in the market also includes training, customer services, documentation, and other support activities associated with the product. Simple products mean less time and effort in those activities, which convert into lower product cost and higher profit margin. At the same time, it is easier to find the resources and talent for simpler products. However, a company may lose competitive advantage from technical point-of-view. Therefore, a right balance should be maintained between these factors. Stakeholders in the controls industry are well-informed about PID and simple FSM strategies. Other control algorithms, however, represent challenges and thus the opportunities at the same time. Ideally, a product should be designed in such a way that the front-end (user-facing part) should be easy and simple, but the back-end can be complex (yet sophisticated) to preserve competitive advantages and core features of the product. On a similar note, it is also easier to find talent for simple control products or the technologies that had been used for a while in the industry.

- *Computation:* Computational power is another important aspect of controls. Larger is the computational power needed by a controller, the higher is the product cost. Existing ASCs or embedded controllers have restricted computational power and memory. Any advanced controller is likely to deploy an architecture and the algorithms that use high computational power. It means that the control algorithm and architecture should ensure that the existing hardware and infrastructure support the control design. For example, any centralized MPC can be implemented only on a supervisory controller or workstation inside an existing BAS.

- *Data quality:* Data quality is dependent on the type and attributes of data required to execute a control algorithm successfully. This includes frequency of data (change of value (COV) or fixed poling time), accuracy of data, time step of data, and historical data points. Data can be an internal variable, measurements from a sensor, or a user-defined input/parameter. Another important factor that increases the complexity and cost is the data cleansing or data preparation strategy needed for the controller, e.g., high fluctuations in power demand may require to average out the measurements and use the averaged value as the final power measurement.

- *Debugging:* Debugging is a process of identifying the root cause of a problem that had occurred in a controller or a system. Even in a well-tested controller, problems could arise due to several reasons such as controller facing an untested scenario, controller operating outside the allowed range, degradation and changing dynamics conditions because of long periods of operation. Rule-based/FSM local controller is relatively easy to debug as compared to a distributed stochastic MPC controller, which is executing a complex logic. Using the same reasons as mentioned in the learning and training bulleted point, rule-based logic can be easily understood while resolving site problems remotely and conveniently to provide better and fast customer service.

- *Robustness and sensitivity:* Robustness and sensitivity help to understand the effect of uncertainties, disturbances, noises, model parameters, model structures, and other perturbations on the controller performance. It is also valuable to include the controller response to outliers and communication problems such as loss of data packets and timing issues during data transmission from point A to point B. Some issues could be handled using data cleansing and filtering techniques while other issues need to be considered during the controller design. Generally, the sensitivity escalates with increase in data requirements for the same type of control methodology. However, in different methodologies or categories of control algorithms, the sensitivity can be different. For example, MPC is often more sensitive as compared to a simple PI based controller. However, in the MPC category, an MPC controller using only building dynamics is less sensitive than the MPC controller that uses occupancy measurements [19] in addition to building dynamics. Monte Carlo or Bayesian methods are common practical methods to perform such analysis [49, 19]. A few robust and non-linear control algorithms are less sensitive and less robust to the system uncertainties, though their performance may not be better than the calibrated rule-based controllers, especially when the processes are well-known or well-understood.

- *System response and stability:* Response includes the time (and its associated value) needed for the system to reach a certain state. There are different metrics to evaluate the system response such as rise time, settling time, overshoot, delay time, and the final time corresponding to the steady state. Every controller has unique design criteria for a system response. Trade-offs are involved in the performance metrics. For example, in a PI controller, lower rise time leads to higher overshoots, which could be improved by adding the derivative component. However, noise in the system causes even worse controller performance when the derivative component is present. The situation can be improved by adding filters and compensators.

  Similar to the system response criteria, stability in control design should be considered. There are many stability types in terms of theoretical controls such as BIBO (bounded input bounded output), marginal stability, Lyaponuv stability, and input to state stability. In the buildings industry, BIBO stability is often referred to as the stability. Because of physical and actuator saturation/limits in building systems, the controller (or the equilibrium point) does not necessarily go unstable but the system may exhibit oscillating behavior or low performance [45].

- *Reliability and fallback mechanism:* Reliability corresponds to the controller's ability to produce consistent results and performance. If a fault or mechanical failure occurs in the system, the controller should act to bring the system to a safe state. This requires developing a fallback mechanism and a transitioning strategy to switch states from fallback to normal operating state. Reliability and fallback mechanisms are critical for several main reasons: (1) to produce high quality product or service; (2) to preserve brand value by reducing inconsistencies and user complaints; and (3) to reduce equipment, property, and personal damage. Therefore, almost every major equipment in buildings with embedded controllers have built-in safety mechanisms.

- *Developmental time and testing time:* Time to finish the development and testing of a control product depends on several choices made during the process. Testing is discussed later in detail in Chapter 5. Developmental time and testing time are important as they dictate the frequency of product releases in a given period of time. If a controller has long developmental/testing time, the product will lag behind in using latest technologies and tools. Moreover, the company may miss a business opportunity if the opportunity arises in the middle of long development cycle. At the same time, if the company is developing a new control product which is much superior than its existing products, the quality and features of the product will differentiate itself from others despite its long development and testing time.

- *Upgradation:* Updating the existing control software or control algorithms requires several steps raising several key questions as asked below:

  1. Can the controller be upgraded without shutting down the entire system?
  2. How can a control algorithm be upgraded, e.g., through the cloud, USB device, P2P connection (via SC, ASC, or workstation)?
  3. How modular is the control algorithm, i.e., can it be updated for one controller only?
  4. What is the time required to upgrade the entire control system?
  5. What are the consequences of failed upgradation or partial upgradation?
  6. Can it automatically identify the targeted system?
  7. What are the manual tasks involved during an upgradation, e.g., data mapping, renaming devices and points?

  Selection of control algorithm and control architectures decides the available update options and thus the upgradation pathway. A distributed control algorithm that is less sensitive to the neighboring measurements can be easily upgraded because of its modular nature as compared to a centralized controller. Details on deployment and upgradation are discussed in Chapter 5.

- *Supporting tools:* Productivity and efficiency of a product life-cycle is heavily dependent on the software and hardware tools used along the way. Therefore, the design choices in controls should consider those factors, which are typically ignored because an initial investigation is needed to make the right decisions. The idea is that if an active community is ready to share the knowledge and tools that have been accumulated over years, time/effort during the entire development/testing/maintenance process is reduced. A vibrant software community is likely to have frequent updates solving issues/bugs while incorporating latest technologies. This includes selection of control architecture and making key decisions as part of the architecture details such as programming language, open-source modules and libraries, communication technology, number of users in the community and their contribution, visualization tools, monitoring tools, etc.

- *Deviation from last product:* It is useful to decide if the new control product is an incremental change or the step change as compared to the last product. It is recommended to keep the variations minimum while adding sufficient value to the product. Higher variations cause higher investment (time and money) on every front, e.g., changing the control strategy from a PI controller to an MPC controller requires additional development, testing, training, support, and expertise. On the other side of the coin, small incremental changes are not likely to have major breakthrough and disrupt the market.

- *User interface:* Interface for different stakeholders (end-customers, distributors, installers, operators, etc.) is also crucial for a product. A good user interface can facilitate the entire process and increase the acceptance of product in the market. A user interface may not be heavily dependent on a control algorithm, but it is tied to the selection of design choices in certain aspects. A few features of good user interface are simple, easy to use, fast, pleasant, informative, and intuitive. If the control algorithm, such as MPC, requires too many inputs from users, it is very challenging to develop a user interface in contrast to a simple "if-else" control algorithm. Another example is related to a PID controller: Suppose tuning or calibration parameters are entered by its user (technician or operator) manually. In that case, it is better to implement a PI controller than a PID controller if there is slight performance difference between the two. This way the user's manual effort reduces by 33% as the user has to enter or tune only 2 values instead of 3 values. Suppose that there are 100 VAV boxes in an AHU-VAV system and calibrating each gain PID takes 2–5 minutes, using a PI controller instead of PID controller will save the installer 3.3–8.3 hours on just one system.

There are numerous choices associated with selecting the "best" controls. There is no single right or wrong set of characteristics. Instead, they depend on the business and customer needs. Since every characteristic has its own pros and cons, their selection should be such that they add sufficient value to the product/portfolio/business in achieving both immediate and long-term goals. More importantly, ensure that the customers perceive the same value as the company thinks. Furthermore, the control choices should be made to ensure that they are aligned with corporate and marketing strategies. Information provided in this section is intended to help understand the tradeoff and consequences of design choices so that controls/business leaders can make a thoughtful, conscience decisions that are best for your team and company.

> *"Mathematical analysis is as extensive as nature itself; it defines all perceptible relations, measures times, spaces, forces, temperatures."*

Fourier, *Joseph*

Key Takeaways: A Few Points to Remember

1. Control, similar to the brain of a body, is a very important part of the building ecosystem and is being used widely at multiple levels in buildings.

2. Control is a diverse and broad term involving a variety of products such as embedded controller, control algorithms/applications, control servers, gateways, supervisory controllers, full-fledged controllers, control services and accessories.

3. Value proposition of control in buildings is of high impact affecting many stakeholders throughout the entire value chain.

4. Control offers and attractive solutions because of its high involvement, substantial impact, low investment, and faster developmental/release cycle.

5. Control approach is critical in laying out the foundation and plan for the subsequent control steps as the approach sets up the control architecture indicating relationships/connections between key elements.

6. Hierarchical control is most commonly used control approach in building systems because of the heterogeneous nature and functionalities of building components.

7. Many control algorithms exist in building systems and the most common algorithms are PID, bang bang, if-else, rule-based, and FSM control.

8. Other new (or less pervasive) control algorithms in buildings are MPC, adaptive, and optimization-based strategies. AI strategies have also started to emerge, mostly at the cloud/centralized level.

9. Optimal selection of control requires choosing the right architecture, approach, algorithm, platform, and hardware with careful trade-off between different components and choices.

10. Trade-off variables during the control development process are sensing, mathematical modeling, learning and training, computation, data quality, debugging mechanism, robustness and sensitivity, system response, stability, reliability, success rate, development effort, testing time, operation effort, supporting tools, and user interface.

Bibliography

[1] Ecobee. `https://ecobee.com/`. Accessed: 2020-03-22.

[2] Google Nest. `https://nest.com/`. Accessed: 2020-03-22.

[3] PID Controller. `https://people.ece.cornell.edu/land/courses/ece4760/FinalProjects/s2012/fas57_nyp7/Site/pidcontroller.html`. Accessed: 2020-05-02.

[4] *Principles of Economics*. University of Minnesota Libraries, 2016. Accessed: 2020-12-02.

[5] J. Baillieul and T. Samad. *Encyclopedia of Systems and Control*. Encyclopedia of Systems and Control. Springer International Publishing, 2021.

[6] K. Bala, A. Bracy, E. Sirer, and H. Weatherspoon. Finite State Machines. `http://www.cs.cornell.edu/courses/cs3410/2018fa/schedule/slides/07-fsm.pdf`. Accessed: 2020-04-22.

[7] A. Bernaden. Variable air volume control using a finite state machine. In *1999 European Control Conference (ECC)*, pages 31–36, 1999.

[8] Yuanpeng Chen, Yunxiao Shan, Long Chen, Kai Huang, and Dongpu Cao. Optimization of pure pursuit controller based on PID controller and low-pass filter. In *2018 21st International Conference on Intelligent Transportation Systems (ITSC)*, pages 3294–3299. IEEE, 2018.

[9] Ming-Li Chiang and Li-Chen Fu. Adaptive control of switched systems with application to HVAC system. In *2007 IEEE International Conference on Control Applications*, pages 367–372. IEEE, 2007.

[10] Johnson Controls. Energy Performance Contracting: Innovative Approach to Lowering Energy Costs and Cutting Emissions–With Guaranteed Savings. `https://www.johnsoncontrols.com/services-and-support/energy-and-efficiency-solutions/energy-performance-contracting`. Accessed: 2020-08-11.

[11] R.M. Dansereau. Finite State Machines (Chapter VIII). `https://limsk.ece.gatech.edu/course/ece2020/lecs/lec8.pdf`. Accessed: 2020-04-22.

[12] Frank Dellaert. Hierarchical control. `https://www.cc.gatech.edu/~dellaert/07F-Robotics/Schedule_files/02-HierarchicalControl.ppt.pdf`. Accessed: 2020-05-01.

[13] K.J. Åström and Tore Hägglund. A frequency domain method for automatic tuning of simple feedback loops. In *23rd Conference on Decision and Control*, pages 299–304, 01 1985.

[14] K.J. Åström and Tore Hägglund. New tuning methods for PID controllers. In *3rd European Control Conference*, 09 1995.

[15] K.J. Åström and Tore Hägglund. *PID Controllers: Theory, Design and Tuning*. Instrument Society of America, 1995.

[16] William J. Fisk. Health and productivity gains from better indoor environments and their relationship with building energy efficiency. *Annual Review of Energy and the Environment*, 25(1):537–566, 2000.

[17] Siddharth Goyal. *Modeling and control algorithms to improve energy efficiency in buildings*. PhD thesis, University of Florida, 2013.

[18] Siddharth Goyal, Prabir Barooah, and Timothy Middelkoop. Experimental study of occupancy-based control of HVAC zones. *Applied Energy*, 140:75–84, Feburary 2015.

[19] Siddharth Goyal, Herbert Ingley, and Prabir Barooah. Effect of various uncertainties on the performance of occupancy-based optimal control of HVAC zones. In *IEEE Conference on Decision and Control*, pages 7565–7570, December 2012.

[20] Siddharth Goyal, Herbert Ingley, and Prabir Barooah. Zone-level control algorithms based on occupancy information for energy efficient buildings. In *American Control Conference (ACC)*, pages 3063–3068, June 2012.

[21] Siddharth Goyal, Herbert Ingley, and Prabir Barooah. Occupancy-based zone climate control for energy efficient buildings: Complexity vs. performance. *Applied Energy*, 106:209–221, June 2013.

[22] Siddharth Goyal, Weimin Wang, and Michael R Brambley. An agent-based test bed for building controls. In *American Control Conference (ACC), 2016*, pages 1464–1471. IEEE, 2016.

[23] Lv Hongli, Duan Peiyong, Yao Qingmei, Li Hui, and Yang Xiuwen. A novel adaptive energy-efficient controller for the HVAC systems. In *2012 24th Chinese Control and Decision Conference (CCDC)*, pages 1402–1406. IEEE, 2012.

[24] Alex Bernaden III, Gaylon M. Decious, John E. Seem, Kirk H. Drees, Jonathan D. West, and William R. Kuckuk. State machine controller for operating variable air volume terminal units of an environmental control system, U.S. Patent US6219590B1, 1998.

[25] Johnson Controls Inc. Introduction to PRAC (Pattern Recognition Adaptive Control). https://www.johnsoncontrols.com/en_au/-/media/jci/be/australia/air-systems/variable-air-volume-terminals/files/be_pracpatternrecognitionadaptivecontrol_whitepaper.pdf, 2010. Accessed: 2020-08-11.

[26] Srinivas Katipamula and Michael R Brambley. Methods for fault detection, diagnostics, and prognostics for building systems–a review, Part I. *HVAC&R Research*, 11(1):3–25, 2005.

[27] Ranjeet Kumar, Michael J Wenzel, Mohammad N ElBsat, Michael J Risbeck, Kirk H Drees, and Victor M Zavala. Stochastic model predictive control for central HVAC plants. *Journal of Process Control*, 90:1–17, 2020.

[28] Geunho Lee and Nak Young Chong. *Flocking controls for swarms of mobile robots inspired by fish schools*. INTECH Open Access Publisher, 2008.

[29] Daniel Liberzon. Nonlinear and adaptive control. http://liberzon.csl.illinois.edu/teaching/16ece517notes.pdf, 2016. Accessed: 2020-05-01.

[30] Zhiyun Lin, Bruce Francis, and Manfredi Maggiore. *Distributed control and analysis of coupled cell systems*. VDM Publishing, 2008.

[31] Yushen Long, Shuai Liu, Lihua Xie, and Karl Henrik Johansson. A hierarchical distributed MPC for HVAC systems. In *2016 American Control Conference (ACC)*, pages 2385–2390. IEEE, 2016.

[32] Georgios Lymperopoulos and Petros Ioannou. Building temperature regulation in a multi-zone HVAC system using distributed adaptive control. *Energy and Buildings*, page 109825, 2020.

[33] Yuri Mitrishkin and Rodolfo Elias Haber Guerra. Intelligent Hierarchical Control System for Complex Processes–Three Levels Control System. In *ICINCO 2009 - 6th International Conference on Informatics in Control, Automation and Robotics, Proceedings*, volume 1, pages 333–336, 01 2009.

[34] James B Rawlings, Nishith R Patel, Michael J Risbeck, Christos T Maravelias, Michael J Wenzel, and Robert D Turney. Economic MPC and real-time decision making with application to large-scale HVAC energy systems. *Computers & Chemical Engineering*, 114:89–98, 2018.

[35] Kurt Roth, Detlef Westphalen, Michael Feng, and Patricia Llana. Energy impact of commercial building controls and performance diagnostics: Market characterization, energy impact of building faults and energy savings potential. Technical report, U.S. Department of Energy, Nov 2005.

[36] Jeffrey Schein, Steven T Bushby, and Jeffrey R Schein. *A Simulation Study of a Hierarchical, Rule-Based Method for System-Level Fault Detection and Diagnostics in HVAC Systems.* US Department of Commerce, National Institute of Standards and Technology, 2005.

[37] John E. Seem, Gaylon M. Decious, Carol Lomonaco, and Alex Bernaden. Hybrid finite state machine environmental system controller, U.S. Patent US6408228B1, 1997.

[38] Pervez Hameed Shaikh, Nursyarizal Bin Mohd Nor, Perumal Nallagownden, Irraivan Elamvazuthi, and Taib Ibrahim. A review on optimized control systems for building energy and comfort management of smart sustainable buildings. *Renewable and Sustainable Energy Reviews*, 34:409–429, 2014.

[39] Siemens. Adaptive Control for APOGEE Building Automation. https://www.downloads.siemens.com/download-center/Download.aspx?pos=download&fct=getasset&id1=A6V10304965, Feb 2006. Accessed: 2020-06-02.

[40] Amanjeet Singh, M. G. Matt Syal, Sue Grady, and Sinem Korkmaz. Effects of green buildings on employee health and productivity. *American journal of public health*, 100:1665–8, 09 2010.

[41] Servet Soyguder, Mehmet Karakose, and Hasan Alli. Design and simulation of self-tuning PID-type fuzzy adaptive control for an expert HVAC system. *Expert systems with applications*, 36(3):4566–4573, 2009.

[42] John Swigart and Sanjay Lall. *Decentralized Control*, volume 406. Springer, 10 2010.

[43] H. G. Tanner, A. Jadbabaie, and G. J. Pappas. Flocking in fixed and switching networks. *IEEE Transactions on Automatic Control*, 52(5):863–868, 2007.

[44] Trane Technologies. Performance Contracting. https://www.trane.com/commercial/north-america/us/en/controls/energy-services/contracting-and-financing/performance-contracting.html. Accessed: 2020-08-11.

[45] Bhagyashri Telsang, Mohammed Olama, Seddik Djouadi, Jin Dong, and Teja Kuruganti. Stability analysis of model-free control under constrained inputs for control of building HVAC systems. In *2019 American Control Conference (ACC)*, pages 5878–5883. IEEE, 2019.

[46] R.M. Underhill and D.J. Taasevigen. Army re-tuning implementation guides. Technical report, Pacific Northwest National Laboratory, 2019.

[47] Michael J Wenzel, Robert D Turney, and Kirk H Drees. Model predictive control for central plant optimization with thermal energy storage. 2014.

[48] Alison Williams, Barbara Atkinson, Karina Garbesi, Erik Page, and Francis Rubinstein. Lighting controls in commercial buildings. *Leukos*, 8(3):161–180, 2012.

[49] Xiaojing Zhang, Georg Schildbach, David Sturzenegger, and Manfred Morari. Scenario-based MPC for energy-efficient building climate control under weather and occupancy uncertainty. In *2013 European Control Conference (ECC)*, pages 1029–1034. IEEE, 2013.

5

Testing and Deployment

Testing and deployment are two critical areas of any control(s) product development. Industries, as compared to academia or research laboratories, pay much higher attention to both the areas as they set the tone for successful products that are planned to be used for the customers on large scale. This chapter discusses the steps involved in testing and deployment phases for delivering quality control products in buildings.

5.1 Test Bed Design

The primary purpose of testing a control product—regardless of whether it is a new control algorithm only or purely a new hardware controller or a hardware controller with embedded control algorithms—is to ensure that the overall system (or parts of the system) is behaving as expected. Testing the behavior requires a test bed on which the system behavior can be monitored and validated efficiently. The first step of designing a test bed is specifying the behavior of system and the scope of testing, which are determined based on product's type and features. Creating a specific set of requirements helps in defining and restricting the scope of test bed. Below are the examples of a few high-level test bed requirements:

- System and interaction: The test bed should be able to interact with certain types of equipment inside buildings. As an example, the control hardware must interact with VAV boxes, occupancy sensors in the lighting system, and a restful API to access outdoor temperature.

- Control mechanism: The control hardware should be able to execute distributed control algorithms. It means that the hardware must execute local control actions while sharing information with neighbors on a field bus, which has limited bandwidth and is typically slower than the supervisory communication mechanisms.

- Simulation and accessibility: The hardware must be able to connect to a particular simulator for easy accessibility and fast testing of preliminary control algorithms. Choosing a simulator or model in the simulator is another exercise, which has been covered in Chapter 3.

- Data storage: Two purposes of data storage are: (1) accessibility to the data in a user-friendly fashion during test failures and (2) a proof of quality validation that can be shared with the stakeholders internal and external to the organization.

- Users: The number, frequency, and types of users also affect the choices of the test bed design.

Once the requirements are finalized, the next step is to design an architecture. The architecture includes the elements and structure of the test bed representing key relationships between the elements, including the choice of software/hardware tools and the rationale behind such decisions. Details on some generic test bed requirements and a sample architecture are reported in [4]. There are four major types of test bed configurations: (1) Simulation, (2) Hardware-in-loop, (3) Laboratory testing, and (4) Field demonstration, which are explained next.

5.1.1   Simulation

Simulation is the fastest and least expensive (in most situations) way to evaluate the performance of a control algorithm. Simulation is usually the easiest method of quick prototyping. However, depending on the type of simulator or model, simulation option may not always be the easiest. In simulation, a model is used to mimic the real system and the control algorithms are implemented on the model. Since there is no hardware involvement, the sampling time, resolution, accuracy, and other properties can be easily changed. It means that the test can be conducted much faster than the real-time clock. The speed of the simulation testing is dependent on the computer or the server processing power along with the constraints imposed by the simulation tools or software libraries used in the process. Usually, some manual tuning of the models or simulation tools is performed because of any of the following reasons: (1) the simulator/model is not highly accurate, (2) the model is too generic, which means that the model is not optimized for a specific domain or application, and (3) the model is very slow to simulate or mimic the real conditions. Tuning parameters in the process can be obtained from experimental data and domain expertise of a personnel. Since there are multiple heterogeneous systems in a building, simulation tools used for the components are also different. Thus, a co-simulation platform is needed to bridge the simulators, e.g., FNCS [2] and BCVTB [6]. Chapter 3 explains the model selection process and simulation tools used in buildings.

5.1.2 Hardware-in-loop (HIL)

In contrast to pure simulation, the hardware-in-loop configuration comprises of at least one real physical component with the rest of the system simulated. One of the major advantages of HIL is that there is no need to create a model for the hardware component as it is physically present in the system. Results in HIL are more accurate than the simulations alone. Higher is the number of hardware components in HIL, the higher is the accuracy of testing as the system is closer to the real scenario. However, the major disadvantage of HIL is that the overall simulation runtime is dependent on the real-physical component. In short, HIL testing is slower than pure simulation testing. Another disadvantage of HIL is that a separate mechanism needs to be put in place for time synchronization purposes. Two categories of HIL are described next.

5.1.2.1 Controller-in-loop (CIL)

CIL corresponds to the configuration that uses real controllers, but the rest of the system (equipment, sensors, thermal and lighting dynamics, etc.) is still simulated. Hardware interface or API from controllers are coupled with the simulations to communicate information both directions in real-time. It is very hard to simulate the behavior of existing controllers because of several proprietary features owned by manufacturers. In certain situations, analog to digital and digital to analog converters are needed to communicate with controllers. CIL is beneficial to test the supervisory control logic (not at the local level) and to incorporate some of the hardware constraints and in-built safety mechanisms. For example, some residential thermostats do not allow users to change the setpoint faster than 5 minutes, while others accept the change initially but reject the command internally after a few minutes. With CIL, one does not have to need to fully understand or model all the internal behaviors associated with the controller.

5.1.2.2 Equipment-in-loop (EIL)

In the EIL configuration, at least one physical equipment (usually a mechanical or electrical component such as AHU, heat pump, transformer, battery, inverter, light, fire damper, and elevator) is coupled with models or simulators to replicate the system behavior. Other large physical systems such as ducts and building rooms are also incorporated into the EIL system [5]. Physical sensors can also be included into these configurations. This configuration is useful to understand the system behavior of certain control strategies. As compared to the previous configurations, the results are highly accurate and close to the actual systems. Modeling needs are reduced in this configuration. However, EIL is expensive to setup and maintain. Furthermore, only certain tests that correspond to the system setup can be performed. For instance, if a heat pump is the only physical equipment setup in the system, obtaining the results for an RTU or AHU (through simulation or on a real-equipment) is not trivial. Time is another major constraint during EIL testing. Therefore, the experiments should be well designed and well thought-out before they are conducted on an HIL test bed.

5.1.3   Laboratory Testing

Laboratory and HIL testing are very similar to each other. In fact, they are regarded as synonymous words in many situations. The main difference between them is that laboratory testing is done on a real system with modifications that enable easier/faster experiments while minimizing damages to the property and organizational assets. A new control algorithm tested on a small, unoccupied building is a type of laboratory testing. Experiments in an occupied room with the manipulated environment to test the specific behavior of algorithm is another case of laboratory testing. Changing the upper and lower limits of variable frequency drive (VFD) set points to avoid motor damage is also considered as laboratory testing.

5.1.4   Field Demonstration

Field demonstration is the final form of testing before the product is launched and being available to the customers on wide scale. In this testing, almost all the processes such as wiring, installation, and software updates are being followed as they would after the product launch. No manipulations or no special changes are made to the existing system. The purpose of field demonstrations is to find, validate, and accept the testing results of products from realistic situations. Moreover, issues that are hard to find in laboratory testing or HIL testing can be identified through this testing. In the same scenarios, the users are fully informed about the detailed changes or detailed functioning of the product so that they can pay significant attention to the feature corresponding to those changes. The users may not be informed about every change and functionality of the product to obtain an unbiased feedback from them. This way, they will be able to identify issues in the overall system instead of focusing only on the changes.

Note that a testing environment can include all the options described above. One can design a test bed that switches between simulation and HIL depending on the product development stage. However, designing and marinating such a test bed can be quite expensive. Therefore, testing facilities or environments are typically designed considering the existing product, not all future possible products. The main benefit of testing is that a well-tested product exhibits higher quality and less problems as compared to prototypes or proof of concepts.

5.2   Product vs. Prototype vs. POC

There are many stages of products (or its predecessors), which possess varying characteristics and require different level of testing. Product, prototype and proof-of-concept (POC) are mostly found in practice. It is well-known that product is used as the final form or result that is being sold to customers to generate sales and profit. In contrast, POC is the form that is used to test major components or highly-uncertain components from an initial idea or thought. It is never sold or being available to the customers. POC's main purpose is to mitigate risk and test the feasibility of an idea. Outcomes and learning from POC development are used to generate a prototype. Prototype is closer to a product with minimal testing, since all major unknowns and issues are resolved or well-understood during the POC development. Only fine details and few unknowns are remaining, which need to be handled in the prototyping stage. Prototyping sets the detailed path and direction of a product. Technical leaders and few business leaders are involved in this stage. Depending on the prototype completion, the prototype can be made available to selected customers to validate the market and obtain customers' feedback.

POC and prototyping are combined together occasionally to expedite the process. Although POC is a highly uncertain project with high level of exploration and research, POC has the fastest development time due to minimal testing, partial development (only major unknown components), and minimal cross-functional (or external) coordination. Only business leaders and a few technical leaders are involved in the POC stage. However, if the POC is targeting an area that requires substantial change in the existing processes, logistics, and supply-chain management functions, the representatives from these functions may be required. Table 5.1 provides a comparison between a product, prototype, and POC.

5.3   Testing Methodology

Once the type of testing is decided and the test bed is designed, a detailed experimental test plan is needed. Three main levels of testing related to controls are: (1) Module-level, (2) System-level, and (3) Feature-level, which are explained next. One of the primary reasons behind this categorization is to reduce the total number of tests and to enable easy/fast resolution of the defects found during testing. The categorization helps to organize the tests and communicate the test results effectively to different stakeholders.

	Product	Prototype	POC
Purpose	• Sell to customers • Increase revenue • Increase profit	• Vet out fine details • Set path for products • Test market demand • Obtain feedback	• Test only few/major components • Feasibility check • Risk mitigation
Cross-functional Engagement	Involves manufacturing, supply chain, and logistics for all necessary functions	Involves manufacturing, supply chain, and logistics if significant changes from last product	Involves cross-functional leaders if significant changes from last product
Scope	Well-defined	Well-defined	Loosely-defined
Speed	Slow	Fast	Fastest
Deployment Scale	High	Low	Low
Testing	All inclusive	Minimal	Almost none
Number of Participants	High	Med/Low	Low
Exploration Level	Low	Med	High
Uncertainty	Low	Med	High
Targeted Audience	Customers	Technical staff and selected Customers	Internal technical leads and business leaders

TABLE 5.1
Comparison of product, prototype, and proof-of-concept (POC).

5.3.1   Module-level

Module-level is the testing associated with the smallest components of a product. It is also known as component testing or unit testing. These tests are at the lowest level in the testing, but at very highest priority level. The tests are written by the person (or a tester) who developed the component. The idea behind the module-level testing is to test all the fundamental blocks once so that they don't have to be repeated at every other level of testing. The module-level testing can be done much faster than any other form of testing. If the failures in the tests at this level are not handled immediately or properly, they can impact the overall system performance substantially. Module-level testing can be as simple as confirming the data exchange between two sensors or confirming if the temperature is being converted from Celsius to Fahrenheit accurately.

5.3.2 System-level

System level testing is a broad term used to test the overall behavior of a device that the customer may or may not able to observe first-hand. This testing is usually performed to test the complete function of control product. An example of system-level testing for a pure control algorithm product is checking the functionalities of duct static pressure set point. In this test case, the behavior of static pressure set point [5] over time based on multiple conditions (fan speed, zone temperature, outdoor temperature, etc.) is checked against the expected results. Similarly, the behavior of changing the lighting (intensity) based on outdoor conditions and availability of natural light through windows is checked under system-level tests. Of course, it is quite difficult to use laboratory testing for this purpose. Therefore, the system is manipulated to reflect the scenario or a model is used to replicate the system behavior.

5.3.3 Feature-level

Feature-level testing corresponds to the test cases that the end-customer is going to face in real-life. Feature-level testing normally involves many components that are working together to provide complete feature(s) that the customer is going to experience. The results of feature-level testing are so general that they can be communicated with customers or a product manager within an organization. User interface can also be a part of feature-level tests. An example of feature-level test on the smart thermostat is that a user should be able to change the set point every 10 minutes, which results in turning on/off of an appropriate equipment and communicate the results back. This feature-level test consists of several system-level and unit-level tests. A button needs to be pushed up or down on the HMI (Human-Machine Interface) screen by the user. Pushing up or down should change the set point. If the room temperature is greater than the set point for more than certain time and certain amount, the air conditioning mode should change to cooling. Cooling mode should send a binary control signal to the equipment embedded controller. Display the equipment response on the thermostat screen in case of an error. Depending on the system size and scope, one feature-level test could be very large consisting of multiple feature tests.

5.3.4 Other testing

Testing could be organized and classified in many ways. A few other control testing methodologies, with some terminologies more common in software testing, are the following:

- Usability testing: In this testing, only the usability of features from customer's perspective are evaluated. User interface, aesthetic, and response time are being considered in this testing. In the smart thermostat example discussed in Section 5.3.3, the usability test checks if the user is able to change the temperature set point, or if the response time from the thermostat is less than 15 seconds.

- Scalability testing: This testing evaluates the performance or behavior of system when the number of participants increases or decreases by a certain amount. Participants can be users, devices, or equipment in the system. For example, a MPC strategy minimizing the energy of AHU-VAV system may behave completely different when the number of VAV boxes are increased from 10 to 50.

- Robustness testing: The tests in this category ensure the reliability and consistency of the overall system behavior in unexpected situations. If the data is not exchanged between devices because the communication is down or the unreliable data is being transferred, how a control product reacts to such situations is part of robustness testing.

- Regulation/Compliance testing: Regulation testing filters the tests as per specific standard or regulatory code so that the product can be certified against the regulatory requirements. Since BACnet^{TM} is a widespread open communication protocol in building systems, BTL certification is used for the same purposes [1]. In certain states in the US, a product can only be sold only if a set of certain certifications/compliance requirements are met. For instance, Title 24 compliance is needed for both residential and commercial buildings in the state of California.

- Sanity testing: Instead of covering many possible scenarios, sanity tests check only major functions from deployment perspectives. For example, in case of new control algorithm development, it will be checked whether the control software packages are correctly compiled and built so that it can run on certain models of application specific controllers. In some cases, it may check if the equipment can be turned on manually using the control algorithm. Similarly, for a supervisory controller, a sanity test may include that the controller is turned on when connected to a 24 V power source.

- Recovery testing: Behavior of system when recovered from a previous known state or recovery state is evaluated as part of recovery testing. Control calculations are being performed continuously in real-time. In some cases, the current calculations are dependent on historical data or past states. If a control system restarts after a power failure, what would be the expected behavior for different conditions. Similarly, the reference signal (set point) is needed for almost every control algorithm. Ideally, the controller should store all the information. However, accessing or storing a large amount of historical data or state data is not possible. These types of questions are answered through recovery testing.

Design of experiments (DOE) offers another systematic, statistical way to define and conduct tests for proving a hypothesis or determining the behavior/relationships between output and input variables [3].

5.4 Validation Mechanism

The detailed test plan needs to be implemented on the test bed for validation purposes. There are three main mechanisms through which the results of tests can be verified and accepted: (1) Eye-ball, (2) Semi-automated, and (3) Fully-automated.

5.4.1 Eye-ball

Eye-ball validation is the traditional and most primitive form of validation. In this validation, a tester (a person) executes all the functions manually using his/her eyes, hence called eye-ball validation. The functions include changing the inputs, pausing/waiting for the results, calculating the tolerance, and comparing the actual output against the expected output. This type of validation is very expensive because of manual, labor intensive tasks. Eye-ball testing is the quickest and easiest method of validation because no prior preparation (except some basic test plan) is needed to run the tests and check the test results. Furthermore, there is no software development since there is no automation of the tests. This type of validation is usually suitable for POCs and prototypes during research/exploratory stages in which the system behavior is unknown and uncertain. Furthermore, the testing is not repetitive during those stages. If this type of validation is used in a product, the testing process can be quite repetitive, exhausting and prone to errors. Performing such validation demands higher knowledge of the system as each test is being manually executed and visually inspected. For a successful and well-established business, it is expected to lower the total product cost and improve the overall product quality, which can be accomplished by minimizing the eye-ball testing activities. At the same time, it may be challenging to completely remove all the eye-ball validation mechanisms as certain tests are impossible to be automated.

5.4.2 Semi-automated

In semi-automated validation, a number of testing/validation processes are automated. This is most commonly used validation mechanism in building control products and services. Basic approach in this category is that the most repetitive and error prone tasks need to be automated to get the maximum return on investment. Of course, the expertise and capabilities of teams should be evaluated during the process because a software/hardware platform (starting at small scale) is needed to accomplish some sort of automated validation. The development process could be time consuming, especially if the team had not done this before. In general, there is no commercial software or universal platform available that fulfills most of the control product testing requirements. Therefore, a good platform is either developed from scratch or created by leveraging and modifying the existing commercial products. Such validation platforms are developed primarily for internal use, and for customer sites in rare occasions for debugging purposes. Although semi-automated validation is used for products, a minimal level of automation is sometimes incorporated in prototyping, especially when the prototype is being released to a few customers to validate the market or obtain their feedback. Semi-automated validation overcome many limitations of the eye-ball validation. Semi-automation is considered as an intermediate step toward fully-automated testing and validation.

5.4.3   Fully-automated

In fully-automated validation, most of the testing and validation processes are completely automated. It is hard to achieve 100% automation in which there is no human involvement. Software companies lean more toward fully-automated testing and validation while building industry tend to lean toward semi-automated testing and validation mechanisms. However, the trend is changing nowadays with advent of cloud and software technologies in building systems. Fully-automated validation requires the highest fixed and variable costs for the platform. At the same time, this validation offers supreme quality and least time to deploy completely new control products as well as the products that are slightly modified from the last release. Fully-automated validation becomes prevalent in the conditions when there are

1. Numerous repetitive tasks,

2. Frequent testing and validation processes,

3. Frequent software/firmware/hardware updates and releases,

4. Several interconnections between components or modules,

5. Minimal uncertainties in the future direction of product,

6. Requirements on backward compatibility for several releases,

7. Requirements to maintain the product for a long period of time, and

8. High dependencies on external modules such as open software libraries and off-the-shelf hardware.

Fully-automated validation is also applicable to the scenarios when a control algorithm or software is deployed on a variety of equipment or hardware models. Although the scenarios are explicitly stated for fully-automated testing and validation, a number of conditions are also applicable to the semi-automated methodology.

5.5 Validation Criteria

Designing a validation criterion is as important as developing the test cases because inaccurate criteria can lead to poor product quality and high testing cost. Validation criteria (or "acceptance criteria") is a metric used to accept or reject a single test or a group of tests. As discussed earlier, the detailed test plan relies heavily on the use cases and the products. Although validation criteria is also dependent on the products and the purpose of tests, the platform should be flexible enough to incorporate many forms of criterion. This way, the platform will be able to handle not only the current use cases but also the anticipated use cases considering both existing and future products in the pipeline. A few forms or characteristics of validation criteria are briefly discussed here:

- Absolute tolerance: A user is interested in confirming if the difference in expected output and actual output is within a certain limit. Absolute errors can also be used if the direction of output is not important. As an example, one may be interested to check if the zone temperature gets close to 72.0 °F (22.2 °C) when the zone temperature setpoint is changed to 72.0 °F (22.2 °C). It is usually a good idea to not use zero tolerance for certain points. The precision of sensor, the number of significant digits, and analog-to-digital conversions can lead to very minute errors that are not relevant but may end-up with test failures. For instance, if the sensor reads 71.99 °F (22.19 °C), the entire test fails if the tolerance value is put as zero. Therefore, domain expertise is much needed in preparing the test plans and deploying testing procedures.

- Relative tolerance: Relative tolerance is an easy way to accept or reject a test. In the example discussed above, 0.2% tolerance can be used to accomplish a similar goal. However, the dimension changes can affect the results. Moreover, variables with same type or dimension may not have the same relative tolerance. For instance, relative tolerance of chilled water supply temperature should be higher than the relative tolerance of hot water temperature because of difference in their magnitudes.

- Frequency: Frequency is another important attribute to be considered during the validation stage. This includes the frequency of changing an input value, the frequency of reading an output value, and the frequency of test run. Suppose a test includes changing an input value at a certain time instance. However, the platform sends an input value and the input is not changed in the first shot. Frequency of inputs dictates an acceptable number of attempts to change the value. The duration between those attempts is another parameter that depends on the test case. The same is the case for output value. Imagine that the value of zone temperature setpoint is changed at the supervisory controller. When the setpoint value is read at the local controller, the value at the local controller does not match the value at the supervisory controller because it may take a while for the value to propagate down to the local controller. Does it mean that the test failed or does it make sense to read the value again a few times before declaring the test failure? Another important question is: After the first failure, how many times the test must pass immediately to be considered as acceptable pass. These parameters are defined under this broad classification of frequency.

- Time duration: Parameters associated with time are handled and evaluated in this case. Time interval between two attempts after a failure as described in previous bullet point is an example of time duration parameter. Polling time to take certain actions such as reading inputs or writing inputs, waiting time to restart a test after a failure or power cycle are configured and evaluated through time duration parameters. This ensures that the behavior of system is evaluated or manipulated at a pre-specified time instance or within pre-specified time duration. For instance, it does not make sense to read the zone temperature set point every 0.1 s after the value is changed at the supervisory controller because frequent reading interferes with the existing communication, which may further slowdown the other intermediate processes by creating additional communication traffic.

- Past test dependency: Behavior of forthcoming tests based on the failure or success of previous test(s) is evaluated in this situation. Suppose that there is one large test that requires changing inputs at multiple time instances, e.g., evaluating the performance of a PID loop, which is very common in building control algorithms. The output at the current time instance is calculated from previous inputs and responses. In this scenario, failing a test in the middle of the large test will lead to failure in all subsequent tests. Based on the use case and reasoning, the evaluation criteria is determined in advance for the tests that are interlinked to each other.

- Failure behavior: It is a broad categorization in which specific rules and guidelines are provided when a test or test suite fails. This may include many types of failures such as communication failures, reliability checks, outliers, synchronization of tasks, executing actions, and reading values.

- Protocol dependency: Protocol dependency is more of an attribute of the testing and validation platform rather than purely an evaluation criterion. However, in certain scenarios the system behavior corresponding to different protocols is also evaluated. Many companies offer products with dual interfaces, a proprietary communication for their own products and an open protocol to integrate with other company's products. BACnet is most common open communication protocol used in several building components including BASs. If the entire platform and test cases with validation criteria are designed around one specific protocol, it becomes challenging to extend to other products or evaluate their performance against other protocols.

- Individual vs. test suite: There are numerous options to write the same test with same validation criteria. Tests can be written individually or combined together in one large test. Moreover, the tests and validation criteria can also be organized in modules and sub-modules. A little thought ahead to organize the tests can yield high time savings, especially when new tests are written or the existing tests are debugged or rerun.

- Short-term vs. long-term: It is very hard to cover a large number of scenarios in short-duration tests. Therefore, long-terms tests need to be established in parallel to discover unexpected events, e.g., memory leaks. The evaluation criteria and the tests for these scenarios may be very different than other tests. For example, running a long duration test to check that the computational time of an advanced control algorithm (suppose model predictive control) is less than the execution time step. Another example is to check the memory consumption of controller over time because the controllers installed in buildings are expected to run for years without any major problem.

- Variable types: Every variable in the system is not numeric. Other variables are binary, text, enumerations, images, alarms, and videos. Validation criteria for such variables should be handled separately. It is not reasonable to put a tolerance on enumerations, e.g., 3-speed fan can only show high, medium, and low as its value. Imposing a relative or absolute tolerance wastes time and leads to wrong results. These checks and balances improve the overall testing platform.

5.6 Creating a Powerful Testing Infrastructure

Major processes involved in a test plan are summarized in Figure 5.1. The first step in creating an effective test plan contains gathering information from internal stakeholders to finalize the list of requirements. This may require interacting with product managers and technical leads regardless of the product type, product stage (old vs. new), and its associated features. POC may not require any test plan, but a completely new product will require substantial testing effort as compared to an incremental change in the existing product. Initial high-level use cases are defined based on the product, its features, and the targeted timeline for the product development and release.

Such requirements are explained earlier in Section 5.1. The requirements lead to a test bed design for testing the targeted functionality of the product.

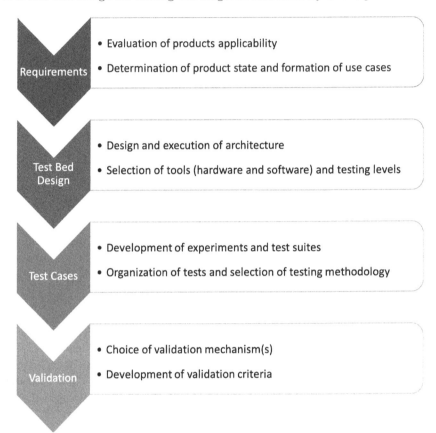

FIGURE 5.1
Overall test plan.

Test bed design comprises of (1) creating an architecture that supports not only the existing requirements but also a few immediate foreseeable needs, (2) selecting a set of software and hardware tools to test the product, and (3) choosing the right mix of simulation and physical components such as devices, controllers, and equipment. With the execution of architecture along with the setting-up of right software and hardware tools, the test bed is developed and is ready to deploy the test experiments. Timeline is critical in making decisions at every stage. For instance, if a completely new product needs to be launched in the next 6–9 months, it is very strenuous to test the product through field demonstrations. Key decisions and rationale behind the decisions should be captured in the architecture document. Section 5.1 provides details on the architecture and types of test bed.

Another crucial step is to develop the detailed test cases and experiments that need to be conducted on the test bed. Usually, many individual tests are conducted to validate the functioning of one single feature of a product. Internal tests that do not directly interact with the users are also defined. This ensures that the product hardware, algorithm, and programming code are well-covered and there are no major gaps in the testing. Multiple testing methodologies and the organization of tests into the methodologies are determined in this step. If it is not possible to implement a specific test, workaround options are explored to design an alternate set of tests that validate the same feature. Testing methodology and test cases are explained in Section 5.3.

The last step in this process is to create an effective validation mechanism and criteria. For the next version of an existing product, new tests that correspond to the new features are added. For such an incremental product, the validation criteria is usually unchanged from the validation criteria of existing product. However, in case of developing a completely new product, a significant amount of brainstorming and investment is needed initially to test the functionality of the product in an automated fashion. Organization of tests into categories and usage is crucial for the products with large test cases/suites. Such organization yields high time savings in the initial testing phase and communicates the results easily to others. Suppose the validation mechanism is fully automated and many features of the product are being developed simultaneously. In that case, there is no need to run all the tests for a small change in one component because running all the tests may take some time, especially if physical hardware or equipment are involved in the testing. The tests related to the change and the tests on the modules that are linked to the components should be sufficient. To accomplish such a goal and reduce the overall development and testing time, the tests should be organized accordingly and the testing platform should be capable of running both individual and bulk tests as per the structuring. In short, the testing platform ought to be flexible. It is mandatory to run the complete suite of tests before releasing the product or merging several components of the product. See Sections 5.4 and 5.5 for details on validation mechanisms and criteria.

It is paramount that most products issues are identified early during the product development because unresolved or unidentified issues have an exponential affect as the issues progress through different stages of a product life-cycle. Figure 5.2 shows the overall business/cost impact of the issues in multiple stages.

As shown in the figure, fixing an issue in the field—when the product is in the hand of a customer—is far worse than fixing the issue during development phase. To fix the issue on a customer site: (1) a patch needs to be developed urgently, (2) technicians need to be trained, (3) technicians need to be sent to

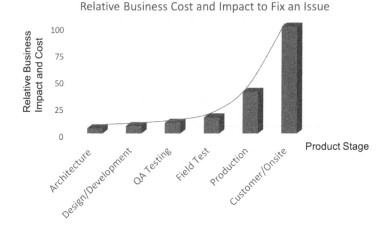

FIGURE 5.2

Cost and business impacts of fixing an issue during different stages of a product.

every building where the product was installed, and (4) analyze the outcome of patch deployment, e.g., follow-up to verify if the patch had resolved the issue. This increases the current and future cost of the product on mass level. Such issues are not only decreasing the product quality but also creating a negative impression on the customers. With the presence of online technologies and the Internet, the information spreads to other potential customers faster than the issue is fixed. As a result, many potential customers are lost and the customers develop a negative perspective or negative image about the brand influencing the business prospects of the company. Therefore, it is necessary to develop and execute a powerful test plan with testing infrastructure to deliver quality products, which can help the business grow.

As discussed throughout this chapter, there are several benefits of testing a product extensively before it hits the market. Figure 5.3 shows the benefits of testing as a function of total testing effort. It is clear from the figure that an initial investment is needed during the product development stage before an organization starts noticing the benefits of testing. As the investment and effort into testing increase, the benefits of testing increase yielding a higher level of automation and platform capability. The benefits increase at much faster rate initially, and slowdown as the extreme/rare/difficult scenarios are incorporated into the testing platform. However, after a certain level of effort, the benefits are almost maxed out. It is recommended to target a definite set of benefits where the return on investment is high while keeping the investment moderate. Usually, it is the area where the curve slope starts to flatten. Although the maximum benefit level is usually not the target, it is important to understand the maximum level because new technologies can either change the maximum level or reduce the cost/time to achieve the maximum level.

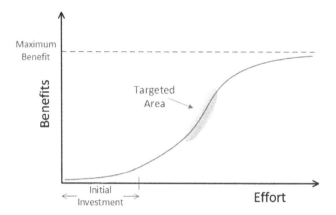

FIGURE 5.3
Benefits of testing as a function of the overall effort.

5.7 Deployment

Now the product is developed and well-tested. The next step is to deploy the product, which comprises packaging, launching, selling, maintaining, tracking, and updating the product. A few aspects of the deployment process relevant to control products in buildings are discussed in this section. Some aspects of this process are decided during the product development stage, while others are settled after the product is developed as summarized in Figure 5.4.

5.7.1 Packaging

Packaging here means combining multiple components together to deliver certain features of a product. In case of a hardware, possible components are sensors, memory chips, connectors, etc. In case of control algorithms or software applications, components are third-party libraries, internal modules, and software tools. Most components are usually decided during the architecture design. However, there are some situations when the decision is altered based on the cost, features, and licensing terms of these components. Licensing, terms and conditions on the usage of third-party components and

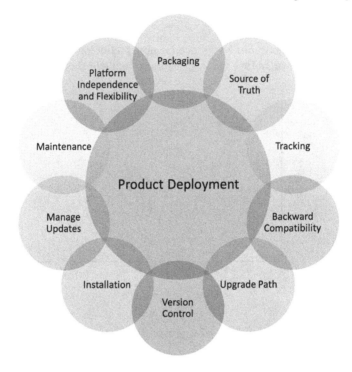

FIGURE 5.4
Steps and key aspects involved during the deployment of a product.

software libraries need to be examined during the architecture design. For example, specific software licenses mandate that any use or update of original software requires the modified software to exhibit the same license as the original software. If a third-party software is open-sourced, the launched software built on top of the third-party software must be open-sourced.

5.7.2 Installation

In this section, the methods of installing a product (hardware or software) on a site are explained. For a control algorithm or software application, there are three options:

- Physical: A CD or a USB is distributed with installation instructions and appropriate credentials for the installation. This is a traditional way of distributing a software and there is minimal tracking or control on further distribution of software, i.e., the same software application can be installed on multiple devices with no licensing restrictions, especially when not connected to the Internet. Sometimes the user has to select the software based on the system configuration and past version of the software. This is a sort of manual process and thus the slowest.

- On-premise: The software along with the instructions is made available to the user via the Internet. The user has to follow a few steps (typically less than the physical case) to install the software.

- Cloud: This is the simplest deployment technique for a user as the software is automatically installed on the system with no action needed from the user. In some applications, users have to manually accept the installation by clicking one or two buttons, but the process is fully automated. Cloud is the most preferred way of deploying a new software (firmware, control algorithm, application) or updating the existing software. However, a good portion of building systems are not connected to the Internet or they do not allow any incoming connections because of security reasons, e.g., critical facilities and small retailers. Therefore, product design and marketing strategy should consider these constraints also.

For a hardware controller, only manual on-site installation is possible. Even though only few other options are available for hardware installation, the product can be developed in such a manner that the installation process is very simple, convenient, and intuitive for the installer. A set of software tools may be created to achieve this objective. As an example, suppose that a company is interested in launching a new thermostat with nicer look. The new thermostat requires an occupancy sensor, which is not present in the last thermostat. The company should aim to design the thermostat in such a way that the installer can snap and replace the front part of the thermostat in each room easily and quickly. In addition, there should be no need to rewire the thermostat and there should be no need to take the back plate out of the wall. Wiring and screwing the back plate into the wall are the two time-consuming steps in the entire process. This way, the company had improved the front part of the thermostat to improve aesthetics and added sensor to the front portion of the thermostat that requires minimal installation.

5.7.3 Automatic Binding

Control algorithms often require many inputs and outputs consuming the measurements from several sensors and user-configured parameters such as zone temperature setpoint or normal working schedule. Deploying control algorithms in these environments is time-consuming, labor intensive, and prone to errors. Automatic binding is a way to find the desired target of inputs and outputs, and make the connections automatically or manually with minimal (or almost no) human intervention.

5.7.4   Tracking

Tracking is a broad term that can be applied in many ways. Tracking of a product constitutes its model number, serial number, and the deployment environment such as product location, system on which the product is installed, system configuration, and installation date. In case of software application or control algorithm, a version number should be sufficient in place of model or serial number. Availability of version number along with a standard version control method helps in (1) prioritizing the next development work, (2) executing automatic or manual updates, and (3) resolving and managing issues. Availability of deployment environment can open new doors of opportunities though the information may not be easily accessible in buildings. Installation date (if available) can generate sales opportunities as new products can be offered to the customers if the product had been installed a while back. However, installation date may not be critical from technical perspective.

5.7.5   Update and Maintain

Update process is mostly applicable to control algorithms or the software associated with control applications. Automatic updates are possible with version control and Internet connectivity. It is highly recommended that the updated control algorithms are backward compatible. It means that the new algorithms should not require any major inputs from user to implement them on their system. Automatic binding is a perfect use case for this. For instance, if a PID loop in a control algorithm is improved with an adaptive PID loop that uses historical data to calculate the gains, the entire algorithm should be easily updated for all the systems, even for the systems that do not capture historical data. Upgrade or update path needs to be clearly defined during the product development stage for both minor and major product releases. For example, suppose there is a simple occupancy-based control algorithm, which changes the zone temperature setpoint based on occupancy in the zone. The algorithm needs to be implemented on the system. In the case of upgrade, the algorithm should not only detect the zone temperature setpoint and occupancy sensor but also make the connections in the control logic automatically so that the building operator does not have to spend time in first finding the desired points and then removing/adding the connections manually.

In fact, with increasing connectivity of BASs to the Internet now a days, small changes or features can be deployed on a site in an automated fashion as soon as the software is developed and tested. In software terminology, this process is referred to as "continuous delivery." Continuous updates reduce the stress of a major release and add value to the products as quickly as possible. Moreover, if there are any problems with the updates, the problem areas can be easily identified and isolated. Meanwhile, the system can be easily rolled back to the previous version. Downtime in maintenance is also reduced via this process. It is important to have a version and algorithm management system. One source of truth is also desired to avoid discrepancies. For example, if there are two versions of control modules at two different places not linked to each other, it is possible that one algorithm might be using one module while the other algorithm may be using the second module. It becomes difficult to maintain and update both the modules at the same time, especially if both are performing the same function. Maintenance also includes removing the outdated modules from the system. Notifying users about upcoming changes (security patches, improvements, removing modules, etc.) is also considered under maintenance.

Key Takeaways: A Few Points to Remember

1. Testing and deployment are two critical pieces for successful products.

2. Testing and deployment strategy decides the quality, the total cost, and the method to deliver new features of a product.

3. Requirements, architecture, test bed design, and validation mechanism are essential for incorporating any useful automation in testing.

4. Automation of testing and deployment is an immediate, straight-forward option to reduce the total cost of a product.

5. Use cases, product features, and product type decide the level of automation needed in testing.

6. Simulation, hardware-in-loop, and field demonstrations are three types of test bed. Requirements, product features, and use cases determine the test bed design, test cases, and validation criteria.

7. The cost and business impact to repair an issue grow exponentially as a function of product life-cycle stage.

8. Automated testing and efficient deployment need initial investment and they should be considered as part of the long-term product strategy.

Bibliography

[1] BA Testing Laboratories. `https://www.bacnetlabs.org/page/btl_` `background`. Accessed: 2020-03-22.

[2] Selim Ciraci, Jeff Daily, Jason Fuller, Andrew Fisher, Laurentiu Marinovici, and Khushbu Agarwal. FNCS: A framework for power system and communication networks co-simulation. In *Proceedings of the Symposium on Theory of Modeling & Simulation-DEVS Integrative*, page 36. Society for Computer Simulation International, 2014.

[3] Jay Dhariwal and Rangan Banerjee. An approach for building design optimization using design of experiments. *Building Simulation*, 10(3):323–336, Jun 2017.

[4] Siddharth Goyal, Weimin Wang, and Michael R Brambley. An agent-based test bed for building controls. In *American Control Conference (ACC), 2016*, pages 1464–1471. IEEE, 2016.

[5] Sen Huang, Weimin Wang, Michael R. Brambley, Siddharth Goyal, and Wangda Zuo. An agent-based hardware-in-the-loop simulation framework for building controls. *Energy and Buildings*, 181:26 – 37, 2018.

[6] Lawrence Berkeley National Laboratory. BCVTB: Building Controls Virtual Test Bed. `https://simulationresearch.lbl.gov/bcvtb`. Accessed: 2015-08-20.

6

Control Use Cases, Artificial Intelligence, and Internet of Things

"Strong reasons make strong actions."

— William Shakespeare

Use cases capture the scenarios and system behavior—from business perspective—in which controls (or parts of controls) can be utilized and linked to specific user needs. Use cases provide a high-level picture to the readers and help them understand the problem statement along with its goals, benefits, and the affected stakeholders in their day-to-day lives. In this chapter, a few such use cases and their connections to the business cases in the buildings industry, not just HVAC systems, are highlighted. As part of digital transformation/solution, AI (Artificial Intelligence), ML (Machine Learning), and IoT (Internet of Things) are trending topics, which have serious connections to controls and may impact the building industry significantly in the upcoming decade. These topics are also discussed in this chapter.

6.1 Control Use Cases

There are hundreds of control use cases involving multiple stakeholders providing diversified benefits. The goal here is to discuss a few use cases and cover a lot of ground. This way the readers can better understand the holistic picture and develop new use cases based on the requirements. These use cases are attempted to provide different types of benefits such as energy, economic, environment, and power grid. Sometimes the use cases are referred to as automated supervisory control functions/applications, which improve the overall functionality of the system. The use cases along with their benefits and beneficiaries are described next.

6.1.1 Precooling and Preheating

Precooling cools down a building during the times when the electricity prices are low. Stored energy in the building structure is utilized at later times when the electricity prices are higher. As an example, buildings can be cooled during nighttime or in the morning before occupants arrive. Buildings have in-built thermal mass, which is utilized to absorb or dissipate energy into the space. Similarly, building can be preheated ahead of time to avoid excessive heating at later stages when electricity prices are high.

Benefits and Stakeholders

Precooling/preheating primary target is to reduce the overall electricity bill for building owners by reducing the peak demand charges and consuming lower energy during the times when electricity prices are high. Since most residences pay flat or tiered electricity pricing, residential buildings are not likely to benefit much from precooling and preheating. Precooling/preheating is not intended to reduce the total site energy usage. In fact, precooling or preheating may increase the total site energy usage in most cases.

6.1.2 Ice/Cold-water Storage System

Significant increase (almost 70%) in high peak energy demand over the past 20 years had gained the attention of researchers and the industry. Buildings play a key role because of their large contribution to both energy and electricity usage. Construction of new power plants is an option to tackle this situation. However, a new power plant requires high investment and raises several environmental concerns such as pollution, reduction of air quality, greenhouse effect, and acid rain [46]. Ice or cold-water storage system comes into play in this situation because they provide a clean and stable source of energy in buildings to reduce the peak demand although they also require some initial capital investment. The use case is simple that ice storage systems make ice at night and dispatches it during the daytime. Similarly, the cold-water storage systems cool the water and store it in a huge tank at night, and dispatch it during the daytime. There are several variations of ice/cold-water storage systems, but the high-level concept is the same. As an example, an additional chiller (glycol-based) is installed to make ice and keep the ice in ice storage tanks at night. During daytime the ice is dispatched through the pipes going through a heat-exchanger, which transfers cold energy to the existing chilled-water entering the existing chillers.

Benefits and Stakeholders

Similar to precooling/preheating, an ice or cold-water storage system reduces the overall electricity bill for building owners by reducing peak demand charges and consuming lower energy during high electricity prices. They are primarily applicable to large commercial buildings claiming 10–20% monetary savings. These systems also increase the total site energy usage.

6.1.3 Duct Static Pressure Reset

Duct static pressure reset uses a control algorithm to optimize the operation of the supply fan at air handling units. The idea is to reduce the fan speed when the entire building has low air flow rate requirements. Instead of running the AHU supply fan at full speed and closing the dampers, the fan should run at lower speed while opening the dampers further at the zone level. This reduces the resistance through the duct and thus reduces the fan energy consumption because of fan dynamics [32]. In this case, instead of directly changing the fan speed, duct static pressure setpoint is changed. Another PI loop controls the fan speed to maintain the duct static pressure setpoint [32]. Trim and response logic is a simple strategy used for resetting the setpoint [37]. However, fan operation can benefit from advanced control strategies.

Benefits and Stakeholders

The main advantage of duct static pressure is the energy savings to building owners or facility management companies to meet their annual energy targets. Utility companies can also benefit if the goal of strategy is to manage the demand and provide other demand response services to the grid.

6.1.4 Optimal Start/Stop

Optimal start is an activity of turning on the system at the right time to save energy while maintaining best comfort for the users. Optimal start is applicable to several building systems and their components such as HVAC, lighting, domestic hot water, refrigerant, and power systems. Optimal stop represents a strategy to turn off the system at the right time to save energy while maintaining comfort levels for its users. For instance, hot-water system can be turned on before people start using the hot water inside buildings, e.g., taking a shower in the morning. However, heating up the water is not instantaneous in most hot-water systems. Therefore, the hot-water system needs to be turned on a few minutes before the water is being used. Determining the right start time may depend on several factors such as outdoor environment conditions, day of the week, equipment response time, etc. Similarly, HVAC systems can be turned on before the occupants enter a building. Figure 6.1 compares the working of an optimal start/stop against the working of a scheduled start/stop algorithm.

In a scheduled start/stop algorithm, the system is turned on at a predefined time regardless of the conditions. In the scheduled start/stop, the building is getting occupied at 8:00 am and the system is turned on at a pre-specified

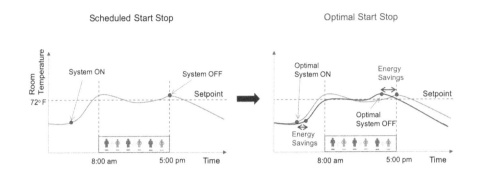

FIGURE 6.1
Schedule start/stop vs. optimal start/stop.

time (let's say 7:00 am) so that the building reaches the desired temperature before the occupants start entering the building, i.e., 8:00 am. It is clear from the figure that the room temperature reaches its setpoint many minutes before 8:00 am. In optimal start/stop, however, the controller determines the optimal start time (let say 7:20 am) so that the room temperature reaches the setpoint exactly (or close enough) at 8:00 am. This way the system is able to save additional energy that was wasted during the first 20 minutes. The same principle is applied for the optimal stop part. Instead of shutting down the system at the scheduled time 5:00 pm, the system is turned off a few minutes before so that the room temperature is still close to the setpoint at 5:00 pm.

Benefits and Stakeholders

The main advantage of optimal start/stop is energy savings to building owners or facility management companies to meet their energy targets. Building operators also reap secondary benefits because they neither have to manually provide starting/stopping schedules nor have to update the schedules as the building degrades or changes over time.

6.1.5   Night Purge Ventilation

As the name indicates, night purge ventilation replaces the stale air with fresh cooler, outdoor air during nighttime. Because of stratification, warm air and pollutants can be trapped in the plenum, roof, and furniture. Some buildings also have sources that generate/absorb high level of heat, water content, and pollutants such as plug loads, laboratory equipment. Night purge ventilation is targeted to remove the heat, humidity, and pollutants and other non-desired particles from indoor areas.

Benefits and Stakeholders

The biggest advantage of night purge ventilation is providing fresher indoor climate to occupants because of removing stale air from the building. Fresher air improves IAQ and thus the health, comfort level, and productivity of occupants. Night purge ventilation also has a partial pre-cooling effect depending on the weather climate and season. It means electricity bills can also be reduced.

6.1.6 Demand Control Ventilation (DCV)

DCV provides ventilation to space based on the real-time requirements in a space instead of supplying a fixed amount of fresh air throughout the day. The most common form of DCV is applied at the building level or floor level, in which CO_2 sensor for each air handler is used as a substitute to determine the amount of fresh outdoor air. Outdoor fan or outdoor damper is controlled to maintain the value of fresh outdoor air. There are many other ways to determine the ventilation demand in a building, e.g., temperature-based models, PIR sensors in individual zones [39].

Benefits and Stakeholders

The main advantage of DCV includes reducing the energy consumed in cooling or heating the excess outdoor air. It means energy savings for building owners. Another benefit is to improve the IAQ through ventilation in case of unexpected events or high-occupancy scenarios, e.g., a town hall meeting or a large training session.

6.1.7 Temperature Setback

Temperature setback changes the temperature setpoints (reference) of several processes in buildings to optimize its operation and reduce energy consumption. There are many processes in different parts of building where the temperature of a liquid or refrigerant is controlled to a certain value; these processes are briefed in the following subsections.

6.1.7.1 Domestic Hot Water Temperature Reset

Temperature of domestic hot water system is reset based on its anticipated usage both in residential and commercial buildings. If many people are planning to use the hot-water for shower in the morning at 7:00 am, the temperature is set higher while the temperature is set to a lower value in other situations, e.g., lower temperature during the nighttime.

6.1.7.2   Zone Temperature Reset

In this case, the zone temperature setpoints are set back during unoccupied times (scheduled or sensed) to save energy. Setback is usually done in the morning before the building gets occupied and in the evening after everyone leaves the building [30].

6.1.7.3   SA Temperature Reset

To avoid simultaneous cooling and heating in an AHU-VAV system, the temperature of conditioned air (or the air supplied by the AHU) is changed to a higher value when the cooling demand is lower in several individual zones. This saves energy both at the cooling coils in the AHU and heating coils at the VAV boxes.

6.1.7.4   Refrigeration/Freeze Temperature Reset

Temperature setpoint of a refrigeration system in commercial buildings or freeze in homes is decreased when the demand is lower than the demand in normal time period, e.g., the freeze is not expected to be actively used for several hours during nighttime. The temperature setpoints are switched back to normal values at other times.

Benefits and Stakeholders

The major benefit of the temperature setback use case is energy reduction, which means lower energy bill for building owners. Secondary benefit, only in some cases, is the improved comfort or improved functioning in unexpected situations. Therefore, the secondary beneficiaries can be occupants or facility mangers.

6.1.8   Daylight Harvesting

Daylight harvesting takes an advantage of daylight to turn off (or dim) the lights in the corresponding lighting zones. Sometimes window blinds and lights are coordinated together to harvest the daylights. As an example, in summer when the days are long, the natural light from the sun can be utilized in the areas closer to windows. The natural light utilization depends on the direction of the sun, time of day, blinds position, and user activity.

Benefits and Stakeholders

Reduced energy consumption is a result of daylight harvesting. It means that the building owners are the primary beneficiaries. If a third party is managing the facility and paying the utility bill, they can also gain from daylight harvesting. Improved lighting, in case window blinds/drapes are controlled, is another benefit to the occupants.

6.1.9   Automatic Lights Shutoff

Lighting shutoff had been starting to be adopted in buildings over the past few years. The basic concept is to turn off the lights automatically when no motion is detected for several minutes in an area. The controls are executed per room or per lighting zone. Lighting shutoff is only possible if the building has not only sensors to detect motion/presence/absence in the area but also the ability to trigger the relay by communicating the information. Sometimes manual entry from occupants is required, e.g., placing door keys at the entrance of hotel rooms to turn on the room lights. New lighting systems have initiated the lighting fixtures with embedded additional sensors to make such use cases easier to implement.

Benefits and Stakeholders

• Benefits: Energy savings and longevity of light bulbs.

• Beneficiaries: Building owners and facility management companies responsible for utility bill payments.

6.1.10   Light Dimming

Light dimming increases control flexibility by modulating the lighting intensity according to the needs. For example, in hallways, it is not recommended to turn off the lights for security reasons even when no one is present for a long duration of time. Another example is on personalized lighting space. If a couple of people are present in the entire open-office space, the lights in other (non-occupied) areas can be dimmed as a function of distance from the occupied space.

Benefits and Stakeholders

Similar to other lighting use cases, light dimming saves energy although replacing the lighting fixtures requires a lot more investment than just replacing high-efficient light bulbs or deploying automatic light shutoff. The beneficiaries are building owners and the facility management companies that are responsible for electricity bill payments.

6.1.11   Occupancy-based Control

Occupancy-based control is a broad category of use cases in which occupants' information (sensor measurements, schedules or manual entries by occupants) is used to control the building systems. The building systems can be controlled together in a coordinated fashion. Occupancy-based control is intended to not only reduce energy consumption but also provide actions that are best for occupants. The best actions can lead to better occupants' experience, better comfort, better IAQ, or better productivity. One example [29] is to use the number of people from occupancy sensor to decide the ventilation rate and their zone temperature set points. Historical occupancy data can be used to predict the occupancy and change the control actions accordingly to provide best comfort and IAQ while saving energy in buildings.

Benefits and Stakeholders

The primary beneficiaries of occupancy-based control are occupants and building owners because of occupants-focused actions and reduction in energy bills. In very specific scenarios, utility companies can also gain from occupancy-based control if the goal is to provide a power grid service based on the occupant information. It is important to keep in mind that occupancy-based control is used at very basic level (primarily using presence/absence for temperature setbacks and lighting) because of lack of sensing and control technologies applied in buildings.

6.1.12   Natural Heating and Cooling

The idea behind this use case is to use natural cooling and natural heating as part of the HVAC controls as much as possible. This can be accomplished by controlling dampers (RA, EA, and OA dampers) and windows whenever outdoor conditions are favorable. For example, in economizer optimization, cold outdoor air is mixed with the return air to minimize the energy consumed at the cooling coils inside an AHU.

Benefits and Stakeholders

- Benefits: Energy savings and better IAQ.

- Beneficiaries: Occupants, building owners, and facility management companies responsible for utility bill payments.

6.1.13   High-traffic Elevator Control

Use case here is to coordinate the operation of elevators to provide best user experience in terms of waiting time, arrival time, and smooth transition. This type of use case is normally applied in high-traffic areas or high-rise buildings such as casinos, hotels, and convention centers. For illustration, suppose that there is large conference at a hotel's convention center. When the session is over at 5:00 pm, all the participants who are staying at the hotel start walking toward the elevators. In this situation, it is important for the elevator control system to estimate the traffic upfront. If elevator control system is aware of the situation, it can bring most of the elevators on the lobby floor to reduce the waiting time for the hotel guests.

Benefits and Stakeholders

- Benefits: Time savings and better occupant experience.

- Beneficiaries: Occupants and facility management. Building owner may also benefit from this use case, e.g., hotel owner attracting future guests or conferences at his/her location.

6.1.14   Fault Detection and Diagnostics (FDD)

FDD has two parts: (1) Fault detection, in which the current (or potential future) issues during the operation of an equipment of system are identified and (2) Fault diagnostics, where the possible root cause(s) of the problem or the recommendations to resolve the problem are provided. A few examples of faults are dirty air filter, crushed duct work, simultaneous heating and cooling, incorrect refrigerant charge (undercharged or overcharged), faulty sensors, water leakage, improper exfiltration or infiltration.

Benefits and Stakeholders

- Benefits: Time savings while debugging issues, improved comfort level, energy savings, and increased equipment life.

- Beneficiaries: Occupants, facility management, building owners, building operators, and technicians.

6.1.15   Building-to-Grid Integration

Building-to-grid integration is a broad set of use cases in which buildings help and contribute to better operation of electric power grid. In these applications, buildings usually respond to a signal/command from the grid. The signal sent from the grid is focused on better management of the overall power grid. The corresponding response from buildings is also aimed to help the power grid in a best possible way while performing their inherent functions and sticking to their main priorities. For instance, the primary job of an HVAC system in a building is to maintain comfortable environment (thermal, IAQ, humidity, etc.) inside the space for its occupants. However, if no one is present in the building, the system has higher flexibility to deviate from its setpoints to help the power grid. Such applications also categorized under "Demand Response" because end-user demand is managed by utilities. There are three ways to manage the demand: (1) direct control, in which specific commands are issued to controllable loads in buildings; (2) indirect control, in which end-users may take an action based on signal from the grid; and (3) market-based control in which automatic negotiations are made in a bid-type fashion with multiple resources competing against each other to set an equilibrium [12]. There are different price schemes for different purposes such as time of use, critical peak price, extreme day price, real-time price [11]. Automated demand response is another way of implementing certain types of demand response events [52]. Below are a few examples of building-to-grid applications:

6.1.15.1   Peak Shaving and Load Shifting

Peak shaving and load shifting are demand side management strategies. Figure 6.2 illustrates the working of peak shaving and load shifting strategies. In peak shaving, the peak load is trimmed to reduce the peak demand charges. This also results in reduced energy consumption but may increase discomfort as some of building systems may not be operational during that time. In load shifting, the load is moved to the time of day when the electricity demand/prices are low. Load shifting is aimed to reduce the total electricity bill although it may increase the total site energy consumption. Precooling and preheating are examples of load shifting control strategies.

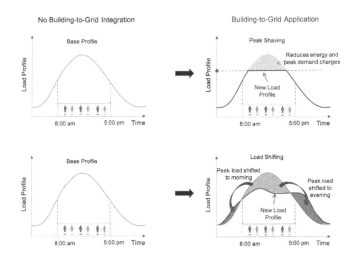

FIGURE 6.2
Illustration of peak shaving and load shifting in a building-to-grid application.

6.1.15.2   Community-level Resource Coordination

In this case, coordination among different end loads provides a grid service at the community level. Figure 6.3 shows the high-level interaction between such components. Community owned generation and storage equipment get activated/used according to the incentive signal sent from the utility company. Computational and simulation tools along with the historical data are used to predict the overall community demand. Community pays minimum price for their electricity charges and manage the internal resources in a way that is beneficial for the overall community and the grid.

6.1.15.3 Frequency Regulation Service

Frequency regulation is a type of ancillary service aimed to increase the transient stability of the system. The goal here is to maintain the frequency as close as possible to 60 Hz. If the frequency changes, the load on the demand side can be accordingly adjusted to bring the frequency back to 60 Hz or close enough. This type of service requires the loads to adjust their behavior very fast, in the order of seconds [38]. In contrast to demand response programs, the average energy consumption at the site level is minimally affected in a frequency regulation service.

6.1.15.4 Demand Response Programs

There have been several demand response programs implemented by utilities and other companies on real-customers. A few programs are listed below:

1. SmartAC^{TM} (Pacific Gas and Electric) [20]: The program uses thermostats and switches to activate the program remotely. In this program, air conditioner cycles at reduced rate (no more than 6 hours a day) to avoid power interruptions. The program is only available from May 1 to October 31, while offering a one-time rebate check ($50) for joining the program along with several other benefits such as free technical support for the AC and free AC checkup.

2. Summer Discount Plan (Southern California Edison) [16]: The program allows the SCE (Southern California Edison) to shut off your A/C's compressor and leaves your home's fan active. Remote-controlled devices at no cost is installed to enable the control. There are two flavors of program: Maximum Savings Cycling and Maximum Comfort Cycling, offering credits up to $140 and $70, respectively. Both programs are available from June 1 to October 1 with override options from end-users.

3. Commercial Demand Reduction (Florida Power and Light) [49]: The program is intended to reduce the peak demand. In this program, a load management device is installed in the building; the device is used to shed a pre-determined load based on an event. The participants are notified in advance about the start time of the event. The participants receive monthly credits to get enrolled into the program.

4. EnergyScout^{TM} for Small Business (Hawaii Electric) [17]: In this program, a device is installed to cycle air-conditioners or turn off water heaters in buildings. The participants receive a monthly check

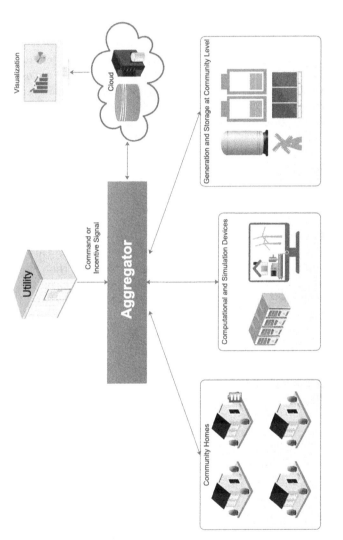

FIGURE 6.3
Building-to-grid application showing integration and coordination of resources at community level.

or credit for each device, $5 for every water heater and $5 for every one ton of air-conditioning system.

5. Capacity Bidding Program (San Diego Gas and Electric) [21]: In the program, the participants pledge to reduce their energy consumption by a certain amount ahead of time. The participants may get notified 2–24 hours ahead to reduce their energy consumption. In return, they receive a payment only when the pledged goal is met. The program is applicable from May 1 to October 31, and it is managed by aggregators and operators.

Benefits and Stakeholders

- Benefits of building-to-grid applications: Reduced utility/electricity bill, reduced energy consumption, and improved management of electric power grid in terms of reliability (e.g., outages) and performance such as avoiding infrastructure costs and higher utilization of efficient generation.

- Beneficiaries: Building owners, facility management companies, utility companies, and generation companies.

6.1.16 Retrofit Measures

Retrofit means replacing an older equipment or device with highly efficient ones. However, in modern sense, the definition of retrofit had been expanded to include new devices, equipment, or provide additional improvements. Retrofits can be coupled with controls because new additions or improvements can open new doors of control opportunities. A few retrofit measures are the following:

- Replacing boilers, chillers, outdoor units, indoor units, domestic hot water systems, heat pumps, fans, with higher efficiency ones to reduce energy consumption.

- Increase insulation of walls, roof and windows by painting, increasing window panes, changing glazing, and using material with higher insulation.

- Add actuation options and increasing control flexibility to the existing equipment, e.g., changing constant speed fans/pumps to variable speed devices, replacing RTUs with heat-pumps, replacing one-mode VRF systems to VRF systems that allow simultaneous heating and cooling in different zones.

- Reducing duct or pipes leakage by installing leak detection devices and filling up the leakage gaps, e.g., Aeroseal [40].

- Replacing lighting bulbs to high efficiency bulbs such as LED to save energy, reduce maintenance cost, and increase life cycle.

- Replacing traditional meters with AMI to allow bi-directional communication between the utility company and end-customer for demand response.

- Upgrading control system (controllers or control software) toward efficient operation while enabling advanced analytics and monitoring on the site.

Benefits and Stakeholders

The biggest pros of taking a retrofit measure are the energy reduction and the operational cost savings for building owners. Occupants may also benefit from retrofits, e.g., replacing traditional thermostats with smart thermostats allows users to change their setpoints remotely. In few cases, depending on the type of retrofit, the facility mangers, building operators, and utility companies may also benefit from retrofits. Most of the retrofit measures require significant capital investment with possible disruption in building operation. The investment decisions need to be carefully evaluated considering non-monetary incentives and the payback time period, which is expected to be in the order of few years.

6.2 Artificial Intelligence (AI) and Machine Learning (ML)

AI and ML are two growing topics and have gained the attention of industries and researchers over the past two decades. The term "Artificial Intelligence" was first introduced by John McCarthy in 1956 although this area had been studied for a while before [5]. AI is a quite generic field encompassing a variety of diversified topics such as natural language processing, cognitive science, perception, image processing, reasoning, logic theory, etc. The entire field originated from the concept that the computers are capable of replicating human brains and human actions, or the thought process in an autonomous fashion; refer to [1, 2, 3] for some general definitions of AI.

AI and ML are often used interchangeably in the industry. However, ML is a subset of AI, dealing with the learning of system behavior while improving/adapting the learning accuracy to the changing conditions. In relation to Chapter 3 on modeling, ML develops a mathematical model (mostly a black box) that updates and improves itself as more data becomes available. Today, AI and ML technologies had been primarily used in the software-dominant industry by companies like Microsoft, Amazon, Baidu, Facebook. However, in the past several years, AI and ML are being adopted in retail, financial, healthcare, entertainment, automotive, and manufacturing industries at different adoption levels for solving multitude of business and customer problems [41]. AI is getting more attention because of improved computational capacity (through distributed computation and faster processors), increased digital devices, increased data storage, development of new AI tools, and advancements in the AI research and algorithms.

In this section, we discuss how the AI and ML techniques can reshape the future of building industry by having a closer collaboration between AI/MI and building systems. It means that AI/MI systems are working together with existing sensors, modeling methods, and control technologies to complement each other and create a wining situation for the buildings' stakeholders.

6.2.1 Role of AI and ML in Building Controls

AI and ML will have a growing role to play in building systems though they have not been embraced yet by the buildings. The most common case of AI and ML found in building is related to security systems that use image processing, visual detection, and natural language processing. If an unexpected event (such as malicious activity, unidentified people, allowed objects) is detected, the authorities are immediately notified for further evaluation. This type of use is found in high-security areas, mission-critical facilities, or secret-sensitive buildings such as labs and research facilities. Alexa [8] and Rumba [34] are examples of path planning, avoiding obstacles, and natural language processing in residential buildings. Figure 6.4 shows a potential timeline of AI/MI including its adoption and use from purely building systems perspective. It is important to note that only few examples are shown to give the readers an idea on what had or what had not been done in the AI area for building systems. Detailed and generic timeline of AI can be found in these articles [6, 55, 10].

Combining the power of AI/ML and controls

Building systems are highly non-linear and complex in nature involving several heterogeneous components. Because of increased unknowns, disturbances, and noise coupled with several changing, uncertain parts that are not measurable, the system becomes very complex to analyze. Therefore, it is almost impossible to have precise and accurate control in those situations. The question becomes: how can AI/ML help in these situations? With the collection of vast amounts of data from different IoT devices and sensors, AI and ML can be used to generate models and use the models to provide a better and faster control solution without necessarily creating an analytical form of the solution. For example, can we model the behavior, style, preferences, and activity level of occupants inside a building? If yes, we can understand their optimal working environment and take the actions that optimize the occupants' productivity.

AI and ML may be able to simplify not only the control development processes by creating on-the-fly models but also provide a simple interface to the users to deploy their own custom control solutions [26]. For example, CFD models require not only deep CFD expertise but also huge computational power to perform any calculations. Furthermore, although the CFD dynamics

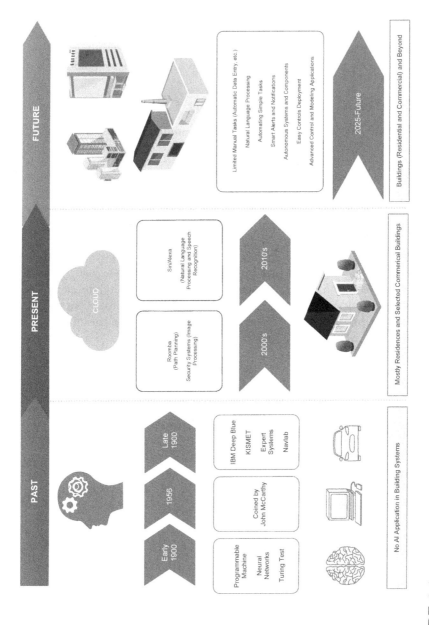

FIGURE 6.4
AI timeline from building systems perspective.

are the same, the calculations are not applicable consistently to every building because of unique building characteristics and several unaccounted factors in the buildings such as opening the doors. In that case, instead of using tuned, offline CFD models, ML models can be generated on-the-fly from real-time data to provide personalized comfort to the occupants.

Controls have a long history and strong foundation for many decades that had been applied in multitude of industries. In controls, there are many tools and techniques available to provide theoretical guarantees of a solution or a system. This is another area of exploration where AI/ML and controls can work together to generate a set of solutions that (1) offer scalability during deployment, (2) provide reliability in terms of consistence performance, (3) provide stability and error bounds on the actions, (4) ensure theoretical guarantees, (5) do not require a large amount of data from very beginning, and (6) explain themselves to certain extent, perhaps partially, rather than being purely a black box [36].

Today, AI and ML is only adopted for a niche market in the buildings industry because of lack of business case justification, e.g., high-development cost as compared to the value generated, lack of awareness, and skill-sets throughout the chain of stakeholders. Its benefits are not comprehended due to lack of sufficient data and proven research cases with field demonstrations on real customer sites. However, in the future, AI and ML technologies are expected to be used to support or take a lead in different building systems, especially when used in conjunction with controls and use of new software technologies. A transitional strategy is needed to increase the adoption of AI and ML techniques. This may include tackling simple scenarios and grabbing low hanging fruits as the first step. Since buildings and the systems inside buildings have a long life-cycle, it is important to start working on fundamental infrastructure and control systems that are installed in the buildings today so that the futuristic buildings are technically capable of deploying the AI and ML technologies. One thing is certain that AI and IoT (discussed in the next section) have to move in parallel to make both of them as reality in the nearby future.

6.3   IoT and Controls

Internet of Things (IoT) had also attracted researchers, academia, and industry over the last decade in attempts to drive innovation, gain competitive advantage, and accelerate business growth. IoT is a system of physical things or objects that can communicate and share information with each other over the Internet with minimal human intervention. Physical things can be sensors, actuators, network devices, or a collection of devices. Physical devices in an IoT system should be uniquely identifiable to enable and manage communication with others. Figure 6.5 shows a high-level schematic of an IoT

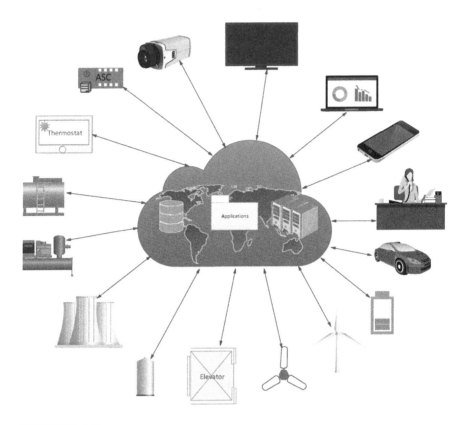

FIGURE 6.5
IoT schematic connecting multiple systems to the cloud; solid line represents a direct or indirect connection (though controllers, gateways, and other devices) to the Internet.

system in which several devices in a building are interacting with each other in real-time. The figure shows a direct connection to the Internet, but the devices can go through multiple levels of devices before it reaches a central place, e.g., cloud. As an example, the sensors can transfer the data to a gateway, which acts as a base station to transfer the data further to a local cell-phone tower that delivers the data finally to the cloud.

Once an IoT device is accessible, new doors of opportunities open to monitor and control the devices based on the data available from several other IoT devices. The concept of IoT is simple to explain but difficult to implement in practice, especially in the building industry because of several technical components and reasons explained later in this section. Therefore, IoT is not as pervasive in buildings as it is in other industries such as software systems and consumables market. However, recent trends in academic research and product development from large and small companies (e.g., IP-based application specific controllers, IP-based cameras, and connected homes) indicate that IoT technology is starting to increase its presence in many building systems [54, 51, 23, 48, 45, 53].

6.3.1 Why Is IoT Attractive Now?

The concept of IoT had been present for several decades although the term was officially introduced by Kevin Ashton in 1999 [35, Chapter 1]. There are many reasons and recent advancements in the technologies that have increased the presence and adoption of IoT in buildings. A few of them are explained below:

- Software Architecture and Interoperability: New enhancements in software architecture and software tools have enabled modularity in the development process. Writing software code and deploying software code on systems had become much easier and faster. Enabling technologies in this area are micro-services, virtualization, and containerization. Use of new open communication mechanisms, communication protocols, and communication physical layers have made it easier to connect heterogeneous IoT devices to the cloud. This way devices can be seamlessly integrated into the existing software infrastructure because of efficient communication protocol conversions.

- Sensing: The cost of sensors had decreased significantly over the past few years. Importantly, the cost of connectivity components in those sensors have been improved substantially as well, e.g., an affordable micro-controller with Wi-Fi connectivity can be found in the range of $1 to $5 [7] instead of $10 to $50 a few years back. These sensors with improved connectivity allow the data to be transmitted from/to the device at much faster rate. The battery life of IoT devices has also increased with advanced hardware and software applications to transmit the right data at appropriate time intervals. Increasing battery life means lower maintenance cost because a technician does not have to replace/charge the batteries frequently. These factors equate to lower ownership cost of an IoT device/system in a building.

- Data Storage: The cost of data storage had also decreased over the past decade. Data storage and data retrieval have become faster and more reliable with increase in data centers, backup technologies, and solid-state drives across the globe.

- Infrastructure Support: The infrastructure to incorporate IoT devices have become increasingly supportive. For example, many new homes come with preinstalled wiring, Ethernet ports, and power connections at different locations to enable the Internet connected devices.

- Data Privacy and Customer Perception: The main component of IoT is the data collected from the devices. Depending on the building type and building location (country or state), customers have varying degrees of requirements in terms of data collection, data storage, data privacy, data security, and data sharing. Overall, the customers have relaxed a certain set of requirements as they are getting educated on the benefits of IoT use cases. Social networking and new technologies to share data anonymously and improve data security have also pacified some privacy concerns.

- Cloud Computing and Networking: Technological advancements in cloud computing and networking have really enabled the use cases that are beneficial for businesses and customers. Uses cases that require large computational power can be addressed with the power of distributed computation, e.g., Service Fabric [44]. During application development, the computational and networking capabilities can be easily utilized and scaled up or down with minimal oversight.

- Application and Use Cases: IoT has not only enabled new use cases that were not possible before but also acted as a catalyst for other technologies. For example, AI and ML applications are feasible only if substantial amounts of data are available. Large amounts of data are possible through IoT. It means that the applications and use cases, which required a long development time historically, can be solved with the help of IoT and AI techniques. IoT use cases and implementation have been increasing in manufacturing supply chain, software, and logistic sectors. Another IoT use case is on big data visualization and monitoring to convey information and convert data into meaningful, actionable items. If more business cases become available and feasible with IoT, it is natural that we will have higher adoption of IoT technologies in the marketplace.

6.3.2   IoT Components

An IoT device has many parts with several decision choices. This section highlights the components that are commonly used in an IoT device. Readers are encouraged to refer to the books [47, 31, 14, 42] for supplemental details.

6.3.2.1   Big Data Handling

IoT devices transmit large amounts of data to the cloud. Some devices have limited storage capacity and thus transfer the data continuously while other devices transmit data at lower frequency as they can store a portion of data locally that is processed further for efficient data handling, e.g., batch processing and transfer of cached data. It may also include the information on how to reroute the data or how to handle the data to avoid inconsistencies and outliers. Similarly, the IoT devices should be able to receive data from the applications running on the cloud to take appropriate actions in a timely fashion. For instance, a cloud application obtains the location of the occupants through their smart phones and sends a "turn-on" command to the lighting system.

Another important part of data handling is the representation of data to indicate what does the data mean and how does one data point relate to other in semantic sense. Semantic models and their frameworks can be very useful in defining things so that machines can understand and interpret things uniquely and consistently across different systems. Open-source metadata schema such as Haystack (simplistic form using tags) [50], Brick [13], SAREF [19] or customized/descriptive schema can be used to create ontologies and semantic models for the same. Haystack have gained some attention over the past several years because of its simplicity and usage. It is expected that formal, sophisticated methods—which create a domain-specific schema that contains vocabulary, relationships, constraints—can enable several practical use cases using knowledge graphs such as faster development of IoT applications [25], easy searching and retrieval of information out of the system [27], and auto-generation of graphics dynamically based on real-time configuration data [28], quick and easy deployment of control, optimization and FDD algorithms [26].

6.3.2.2 Communication Network and Protocol

Physical communication layer is critical for the delivery of data from one point to another. A designer can choose a hard-wired connection (e.g., Ethernet via CAT5) or a wireless connection to the Internet. Despite the high reliability of data transfer in hard-wired connections, a hard-wired connection directly to the Internet is not a viable option in most buildings because of several reasons such as cost, aesthetics, time, labor, etc. Wireless connection comes in two major forms: (1) short-range, e.g., Bluetooth, Zigbee, Wi-Fi, and Z-wave; (2) long-range, e.g., 2G, 3G, 4G, 5G, LoRa, Weightless, NB-IoT, and SigFox. These technologies offer trade-offs in terms of transmission range (or coverage), power, cost, proprietary/openness, and data transfer rate. Selection of connectivity depends on the system architecture and intended use of the device. For example, ASCs of VAV boxes are typically located in ceiling. NFC (near-field communication) alone is not a good option after the controllers are installed as a technician needs to climb up to configure the controllers. However, NFC can save the technician time if the controllers can be setup before they are installed. Furthermore, technicians can setup the controllers without powering them up. Similarly, Bluetooth may not be a good option for IoT RTUs because of their range limitations unless a base station or gateway is installed closer to the unit.

Communication protocols are used to understand the messages that are being transmitted to/from an IoT device. Buildings have diverse range of communication protocols at different levels. For example, in HVAC and lighting systems, the common protocols at the controller, sensor, and actuator levels are BACnet^{TM}, Modbus, Lonworks, or proprietary. It means that the gateway is mostly not IP-based. To achieve high level of interoperability, the IoT devices should be able to support the IP or connect to a gateway that converts the messages to IP. The architecture of building control system and products offered by companies are part of major decision factors. For instance, suppose a company manufactures a thermostat and a heat pump for residential systems which are sold as a single system. In that case, the company does not need to support IoT for both equipment and thermostat. Instead, it can offer an IoT solution for the thermostat and pull-in necessary data from the equipment.

6.3.2.3   Device Management

- Discovery: The first part is to discover the device in the allotted time interval.

- Authentication and Handshaking: The next step is to ensure that only selected and authorized users are allowed to interact with the device. This is part of the cybersecurity practices and protocols associated with the devices. It includes checking the licensing file and making necessary updates to the firmware or software on the device.

- Access Control: Once the device credentials are verified and device is registered with a unique ID in the system, the user can change the configurable parameters (if needed), e.g., authentication to other users, data storage settings, removal of the device from the system, rediscovery of device.

- Data Access: Now the device is ready to transfer data to the Internet directly or through gateways.

6.3.2.4   Platform

Platform means both the hardware and software components of an IoT device to fulfill its operational functionalities. The hardware side includes the architecture, selection of microprocessors, computing (flash memory etc.), digital interfaces, input/output ports, etc. Software decisions include the operating system, firmware to interact with physical objects, startup applications and run-time applications. Software and hardware requirements for half of the times go hand-in-hand with each other. For example, running an Ubuntu OS on Arduino micro-controllers may not be possible. Startup and run-time applications are also dependent on the purpose and usage of an IoT device, e.g., gateway or embedded system. It may include decisions on what applications should be running natively on the IoT device vs. what applications (or what part of applications) will be running on the cloud. Supporting tools are also important to decide the platform requirements for developing and launching applications on the device. A plenty of IoT platforms—mostly from software giants—along with a variety of supporting tools and third-party integration tools exist in the market, e.g., Amazon

AWS IoT [9], Microsoft Azure IoT [43], IBM Watson IoT [33], Google Cloud IoT [24], and GE Predix [22].

6.3.2.5   Data Privacy and Security

Data privacy and security are two biggest issues that need to be tackled continuously in the IoT world. Local and federal agencies (e.g., Federal Trade Commission, California Consumer Privacy Act (CCPA)) watch company policies and impose restrictions on the data collected from the users [15, 18].

Hackers find new ways and gaps in the security systems to hack into the corporate systems through loose ends. IoT can be a loose end if proper security measures are not put in place, which leads to stealing of information and misusing the information for personal or corporate gains. Target hacking through its HVAC system in 2013 is a big case of intrusion [4]. Below are a few measures or techniques (not comprehensive) that can be deployed in an IoT infrastructure:

1. Safeguard against data/query injection
2. Multi-factor authentication
3. Opening only certain ports
4. Disabling or putting restrictions on inactive accounts
5. Applying frequenting security-updates
6. Creating separate VLAN (Virtual LAN) and VPN (Virtual Private Network) networks with incoming and outgoing constraints
7. Changing passwords frequently
8. Changing default passwords
9. Creating strong passwords
10. Applying containerization of programs to limit the accessibility

Similarly, data privacy is a growing concern in the community as stakeholders are getting more educated on the use or impact of data collected from their systems. Transparent communication, clear and concise disclosures, and system flexibility in the system help in establishing trust with the users, especially in the early stages of a project.

6.3.2.6   Applications and Use Cases

Application and business use cases are the two main components of an IoT system. They decide the capabilities of an IoT system in terms of what the system can accomplish and how the accomplished tasks are beneficial for the users. Use cases can be as simple as forwarding the sensor data to the cloud for monitoring purposes or as complex as controlling an equipment using AI techniques. The next section describes such applications at the business level.

6.3.3   IoT Value and Opportunities in Buildings

Building systems and components have been traditionally designed to work as a stand-alone system with minimal input from external sources. The main thinking behind this is to provide reliability in case of malfunctions or failures in other parts of the system. As the technologies started to advance, the concept of BMS and BAS initiated with focus on data monitoring, storage, and visualization. Now BMS systems are quite complex with distributed architectures and several capabilities to control the devices. However, BMS technologies are still way behind the state-of-art technologies implemented in software, networking, sensing, data management, and control area in other industries. A simple AHU-VAV system alone in a small commercial building can generate millions of data values in a day, even when the data is collected at 1-minute interval. Imagine the scenario when every building system and component (including accessory and sensors) behaves as an IoT device, the data collected from the devices can be used for the cases that have not been imagined yet. In many cases, IoT ecosystem (devices and the data collected from the devices) act as an "enabler" to open the new doors of opportunities. It means that in most situations, the data is useless by itself unless it is used by new or existing software/control applications to add value in a product for its users because it is impossible to fathom such amount of raw data by a human in a meaningful way. Below are the high-level scenarios indicating the value and new opportunities in buildings through IoT.

1. New business models: IoT data will drive new business models in the industry. When the data is available in standard form that can be consumed by others, new technology market players will enter the market. Instead of selling equipment and controller in silos, services-based business may emerge that can reshape the entire industry. In a recent global survey of 779 executives, the most common answer from the executives was "IoT will unlock new revenue opportunities from existing products/services" [47, Chapter 1].

2. Efficient control architectures: Reliability and availability of IoT devices allow new control architectures and control approaches that were not possible traditionally. Control architectures will transition from hierarchical to distributed structures with central control done via the cloud. These architectures can lead to efficient workflow of processes and reduction of control-hardware in buildings. For example, instead of using costly supervisory or application-specific controllers, low-cost controllers can be installed with computation distributed on the cloud. However, the supervisory controllers should have sufficient memory and hardware constraints to run the bare-minimum equipment safely in case the system disconnects from the cloud.

3. Insights into system behavior: Large data collected from an IoT infrastructure can be used to capture and study the system behavior providing further insights. This data can be used to approve/disprove the operational claims that a product/company had promised. Data can be used as a part of certification processes mandated by some state agencies.

4. Remote services: Accessing the data remotely and using the data to detect and resolve problems in advance is an example of remote service. By doing so, both users and the company save time as the number of technical visits can be reduced. The system can be tuned manually or automatically to optimize its behavior. Consulting firms can provide guidance to the building owners. Remote services also allow third-parties to better manage the equipment inside a building.

5. Monitoring and visualization: When the data becomes really large, it is important to develop the monitoring and visualizing tools so that users (occupants, buildings operators, remote service providers) can comprehend the information easily and convert it into actionable items. For instance, applying FDD algorithms can generate thousands and thousands of alarms in a simple system. Building operators will have to spend hours and days to go through each alarm. Monitoring and visualization techniques can simplify this process quite a bit.

6. Better historical tracking: To addresses safety and security concerns, it is easier to track the activities of a user, e.g., who is the last person that issued a certain manual command to the system, and when/what data was accessed from the system. Furthermore, historical tracking of data can be used in performance contracting projects, in which a company gets paid to save a certain amount of energy over time.

7. Low-cost installation: The installation cost can be reduced if the devices are able to connect directly or through gateways to the cloud. Electricians do not have to lay wires to make physical connections between the devices reducing both material and labor costs.

8. Better mathematical modeling: Mathematical modeling is essential for three main reasons: (1) to understand the behavior of an equipment or system, (2) to simulate the behavior to develop control algorithms, and (3) to use the model in a control application. Every equipment or advanced controls company spends a quite bit of resources in the modeling. With IoT devices, the data can be easily collected from the systems. Data gathered from diversified environmental conditions can yield better mathematical models in short period of time. Data can be either shared with third parties

for further development or made publicly available to advance the development process.

9. Advanced control applications: With huge amount of new information available, new control applications can be developed. A couple of examples are model-based control and AI applications as discussed in Section 6.2.

6.3.4   IoT Challenges in Buildings

IoT is a growing area of interest in buildings. While converting all (or most) of building systems into IoT systems offer several benefits, there are several challenges—technical and non-technical in addition to the ones mentioned before—associated with the adoption of IoT technologies in the buildings. Overcoming the challenges will certainly help address the concerns of multiple stakeholders. The challenges are highlighted next:

- Lack of skill-sets (both at technical and management level) as the IoT skill-sets are quite different from those needed for the development of building system components.

- Low education and training level as most people may face difficulty in understanding the use of IoT technologies.

- Insufficient infrastructure in the control companies to quickly develop quality IoT solutions, launch them, and stay up-to-date with them.

- Lack of awareness on the value of IoT technologies to improve the entire ecosystem, instead of providing value for just individual users.

- Conflicts between IT and OT systems because of the ownership ambiguity, e.g., who is responsible for maintaining the system, providing networking tools/support, taking liability in case of security preaches, or managing the identities of devices to avoid address clashes.

- High cost of total IoT system because of limited infrastructure support, e.g., in case of IoT sensors such as IAQ and temperature, there are two options: wired Ethernet cable and wireless. Ethernet cables are not available throughout the buildings, and thus incur additional cost when added to the buildings. Wi-Fi is a commonly-used technology to connect the devices wirelessly to the Internet. However, Wi-Fi consumes a lot of power. It means that sensors may need to either provide a large battery-pack increasing the size of the sensor or be changed frequently increasing the maintenance cost.

- Limited regulation constraints/requirements by local and state agencies to drive interoperability between the devices, especially in building sector except the security and IT systems.

- Rough/unavailable/unclear corporate or facility processes inside most buildings for setting up a new IoT device, e.g., several approvals may be needed to add a non-familiar IoT device to the network for security reasons.

- Internet inaccessibility or unavailability at certain locations because of geographic constraints or company policies.

- Inadequate and unaffordable battery technology despite the improvements in the battery technology and reduction in the battery prices over the past decade.

- Concerns over growing electronic waste as electronic devices need to be replaced frequently; most of the devices (or its components) are not available for reuse.

- Insufficient environment or eco-friendly technologies, e.g., discarding batteries and their hazards.

- Skepticism about the reliability, security and scalability of IoT devices, e.g., a common question arises: will the system work if it is momentarily disconnected from the Internet?

Key Takeaways: A Few Points to Remember

1. With more than hundreds of control scenarios that can provide business value to the building stakeholders, advanced control promises tremendous number of opportunities in this sector.

2. Control use cases discussed in this chapter offer benefits for occupants, utility companies, facility owners, building operators, and facility management companies by reducing energy, lowering energy bills, offering better comfort, healthier indoor climate, and longer equipment life.

3. A few use cases are precooling/preheating, reset and setback strategies, optimal operation, DCV, daylight harvesting, natural heating/cooling, high-traffic elevator control, FDD, retrofit measures, and building-to-grid applications.

4. AI corresponds to the science and technologies that are capable of mimicking human brains, human actions, and their thought process in an autonomous fashion.

5. AI is a generic field encompassing a variety of diversified topics such as natural language processing, cognitive science, perception, image processing, reasoning, logic theory, etc.

6. Machine learning (ML) is a subset of AI, dealing with the learning of system behavior of systems while improving/adapting to the changing conditions.

7. IoT is a system of physical things or objects that can communicate and share information with each other over the Internet with minimal human intervention.

8. AI/ML/IoT are growing in the building sector because of advancements in the AL/ML/IoT technologies and reduction in their cost such as computing, storage, sensing, infrastructure, networking, applications, and use cases.

9. Several challenges (technical, educational, cost, ownership, regulatory, etc.) need to be overcome to make AL and IoT as a reality on large scale in buildings.

10. Combining the AI/ML/IoT with controls can provide new and attractive opportunities for both businesses and customers such as efficient architectures, new business models, better mathematical modeling, precise/robust control, lower ownership cost, and advanced control services and applications.

Bibliography

[1] Artificial intelligence. https://www.merriam-webster.com/dictionary/artificial%20intelligence. Accessed: 2020-05-01.

[2] Artificial intelligence. https://www.britannica.com/technology/artificial-intelligence. Accessed: 2020-05-01.

[3] Artificial intelligence. https://www.oxfordreference.com/view/10.1093/oi/authority.20110803095426960. Accessed: 2020-05-01.

[4] Target hackers broke in via HVAC company. https://krebsonsecurity.com/2014/02/target-hackers-broke-in-via-hvac-company/. Accessed: 2020-06-23.

[5] The history of artificial intelligence. https://courses.cs.washington.

edu/courses/csep590/06au/projects/history-ai.pdf,   2006.
Accessed: 2020-05-01.

[6] What's The Big Data? https://whatsthebigdata.com/2017/05/08/
timeline-of-ai-and-robotics/, 2017. Accessed: 2020-06-01.

[7] Olubiyi Akintade, Thomas Yesufu, and L. Kehinde.   Development
of an MQTT-based IoT architecture for energy-efficient and low-cost
applications. pages 27–35, 06 2019.

[8] Amazon.   Amazon Echo & Alexa Devices.   https://www.amazon.
com/smart-home-devices/b?ie=UTF8&node=9818047011.   Accessed:
2020-06-01.

[9] Amazon.   AWS IoT.   https://aws.amazon.com/iot/.   Accessed:
2020-06-29.

[10] Rockwell Anyoha.   Can Machines Think?   http://sitn.hms.
harvard.edu/flash/2017/history-artificial-intelligence/,
2017. Accessed: 2020-06-01.

[11] Mina Badtke-Berkow, Michael Centore, Kristina Mohlin, and Beia Spiller.
A primer on time-variant electricity pricing.   *Environmental Defense
Fund*, 2015.

[12] Sahand Behboodi, David P Chassin, Ned Djilali, and Curran Crawford.
Transactive control of fast-acting demand response based on thermostatic
loads in real-time retail electricity markets.   *Applied Energy*,
210:1310–1320, 2018.

[13] Brick Schema. Brick: A uniform metadata schema for buildings. https:
//brickschema.org/. Accessed: 2020-06-02.

[14] J. Davies and C. Fortuna. *The Internet of Things: From Data to Insight.*
Wiley, 2020.

[15] Earl Duby.   Data privacy in the IoT age: 4 steps for reducing risk.
https://www.csoonline.com/article/3434079/data-privacy-
in-the-iot-age-4-steps-for-reducing-risk.html,   Aug   2019.
Accessed: 2020-06-23.

[16] Southern California Edison.   Summer Discount Plan.   https://www.
sce.com/residential/rebates-savings/summer-discount-plan.
Accessed: 2020-06-02.

[17] Hawaii Electric.   Small Business Direct Load Control Program
(SBDLC).   https://www.hawaiianelectric.com/products-and-
services/demand-response/business-solutions.   Accessed:
2020-06-02.

[18] EPIC. EPIC - Internet of Things (IoT) . `https://epic.org/privacy/internet/iot/`. Accessed: 2020-06-23.

[19] ETSI.   Smart   Appliances   REFerence   (SAREF)   Ontology   .   `https://sites.google.com/site/smartappliancesproject/ontologies/reference-ontology`. Accessed: 2020-06-02.

[20] Pacific Gas and Electric.   SmartAC^{TM} frequently asked questions. `https://www.pge.com/en_US/residential/save-energy-money/savings-solutions-and-rebates/smart-ac/program-faq/smartac-program-faq.page`. Accessed: 2020-06-02.

[21] San Diego Gas and Electric.   Save money with the Capacity Bidding Program.   `https://www.sdge.com/businesses/savings-center/energy-management-programs/demand-response/capacity-bidding-program`. Accessed: 2020-06-02.

[22] GE. Predix Platform. `https://www.ge.com/digital/iiot-platform`. Accessed: 2020-06-29.

[23] Hemant Ghayvat, S.C. Mukhopadhyay, Xiang Gui, and Nagender Suryadevara. WSN- and IOT-Based Smart Homes and Their Extension to Smart Buildings. *Sensors (Basel, Switzerland)*, 15:10350–79, 05 2015.

[24] Google. Google Cloud IoT. `https://cloud.google.com/solutions/iot`. Accessed: 2020-06-29.

[25] Siddharth Goyal.   Building System with Semantic Modeling based Configuration and Deployment of Building Applications, U.S. Patent Application No: 16/379646, 2019.

[26] Siddharth Goyal.   Building System with Semantic Modeling based Custom Logic Generation, U.S. Patent Application No: 16/379652, 2019.

[27] Siddharth Goyal.   Building System with Semantic Modeling based Searching, U.S. Patent Application No: 16/379661, 2019.

[28] Siddharth Goyal. Building System with Semantic Modeling based User Interface Graphics Visualization Generalization, U.S. Patent Application No: 16/379666, 2019.

[29] Siddharth Goyal, Prabir Barooah, and Timothy Middelkoop. Experimental study of occupancy-based control of HVAC zones. *Applied Energy*, 140:75–84, Feburary 2015.

[30] Wei Guo and Darin W Nutter. Setback and setup temperature analysis for a classic double-corridor classroom building. *Energy and buildings*, 42(2):189–197, 2010.

[31] J. Holler, V. Tsiatsis, C. Mulligan, S. Karnouskos, S. Avesand, and D. Boyle. *Internet of Things*. Elsevier Science, 2014.

[32] Sen Huang, Weimin Wang, Michael R. Brambley, Siddharth Goyal, and Wangda Zuo. An agent-based hardware-in-the-loop simulation framework for building controls. *Energy and Buildings*, 181:26 – 37, 2018.

[33] IBM. Securely connect, manage and analyze IoT data with Watson IoT Platform. `https://www.ibm.com/internet-of-things/solutions/iot-platform/watson-iot-platform`. Accessed: 2020-06-29.

[34] iRobot. Roomba Robot Vacuums. `https://www.irobot.com/roomba`. Accessed: 2020-06-01.

[35] J.Y. Khan and M.R. Yuce. *Internet of Things (IoT): Systems and Applications*. Jenny Stanford Publishing, 2019.

[36] Pramod Khargonekar and Munther Dahleh. Advancing systems and control research in the era of ML and AI. *Annual Reviews in Control*, 45, 04 2018.

[37] Sarah Koehler and Francesco Borrelli. Building temperature distributed control via explicit MPC and "Trim and Respond" methods. In *2013 European Control Conference (ECC)*, pages 4334–4339. IEEE, 2013.

[38] Yashen Lin, Prabir Barooah, Sean Meyn, and Timothy Middelkoop. Experimental evaluation of frequency regulation from commercial building HVAC systems. *IEEE Transactions on Smart Grid*, 6(2):776–783, 2015.

[39] Guopeng Liu, Aravind R Dasu, and Jian Zhang. Review of literature on terminal box control, occupancy sensing technology and multi-zone demand control ventilation (DCV). Technical report, Pacific Northwest National Lab.(PNNL), Richland, WA (United States), 2012.

[40] Aeroseal LLC. Aeroseal. `http://aeroseal.com/`. Accessed: 2020-04-02.

[41] B. Marr and M. Ward. *Artificial Intelligence in Practice: How 50 Successful Companies Used AI and Machine Learning to Solve Problems*. Wiley, 2019.

[42] A. McEwen and H. Cassimally. *Designing the Internet of Things*. Wiley, 2013.

[43] Microsoft. Azure IoT. `https://azure.microsoft.com/en-us/overview/iot/`. Accessed: 2020-06-29.

[44] Microsoft. Overview of Azure Service Fabric. `https://docs.microsoft.com/en-us/azure/service-fabric/service-fabric-overview`. Accessed: 2020-08-11.

[45] Daniel Minoli, Kazem Sohraby, and Benedict Occhiogrosso. IoT Considerations, Requirements, and Architectures for Smart Buildings–Energy Optimization and Next Generation Building Management Systems. *IEEE Internet of Things Journal*, PP:1–1, 01 2017.

[46] Volker A Mohnen. The challenge of acid rain. *Scientific American*, 259(2):30–39, 1988.

[47] S.C. Mukhopadhyay. *Internet of Things: Challenges and Opportunities.* Smart Sensors, Measurement and Instrumentation. Springer International Publishing, 2014.

[48] Andreas Plageras, Kostas Psannis, Christos Stergiou, Haoxiang Wang, and B B Gupta. Efficient IoT-based Sensor BIG Data Collection-Processing and Analysis in Smart Buildings. *Future Generation Computer Systems*, 82, 10 2017.

[49] Florida Power and Light. Demand Response Program. https://www.fpl.com/business/save/programs/demand-response.html. Accessed: 2020-06-02.

[50] Project Haystack. Haystack. https://project-haystack.org/. Accessed: 2020-06-02.

[51] A. Rajith, S. Soki, and M. Hiroshi. Real-time optimized HVAC control system on top of an IoT framework. In *2018 Third International Conference on Fog and Mobile Edge Computing (FMEC)*, pages 181–186, 2018.

[52] Tariq Samad, Edward Koch, and Petr Stluka. Automated demand response for smart buildings and microgrids: The state of the practice and research challenges. *Proceedings of the IEEE*, 104(4):726–744, 2016.

[53] Jordi Serra, David Pubill, Angelos Antonopoulos, and Christos V. Verikoukis. Smart HVAC Control in IoT: Energy Consumption Minimization with User Comfort Constraints. *The Scientific World Journal*, 2014.

[54] Tacklim Lee, Seonki Jeon, Dongjun Kang, Lee Won Park, and Sehyun Park. Design and implementation of intelligent HVAC system based on IoT and Bigdata platform. In *2017 IEEE International Conference on Consumer Electronics (ICCE)*, pages 398–399, 2017.

[55] The University Of Queensland, Queensland Brain Institute. History of Artificial Intelligence. https://qbi.uq.edu.au/brain/intelligent-machines/history-artificial-intelligence. Accessed: 2020-06-01.

7

Stakeholders and Channels

"Simplicity is the ultimate sophistication."

— Leonardo da Vinci

7.1   Stakeholders

Understanding the roles and responsibilities of different stakeholders in buildings is one of the most challenging, yet very crucial, tasks. These stakeholders are involved at multiple levels in different stages of building life-cycle processes. Design of a successful control product needs to consider the perspective of the stakeholders although it is very difficult to develop a product that addresses the concerns of every stakeholder. More the number of stakeholders, the higher is the effort needed to understand the behavior of entire ecosystem and develop a thriving product. Typically, products managers, marketing managers, and other business leaders are well-informed about the processes and the role of these stakeholders. However, engineers and control developers should also be aware of such details because they are making the product design/development decisions on day-to-day basis. This increases the chances of a control product to be widely accepted and adopted in the market. Figure 7.1 shows a few prominent stakeholders in building systems. Details on the stakeholders are provided next.

7.1.1   Occupant

Occupants are one of the top stakeholders in the list. Building systems affect occupants the most as occupants are the primary consumers of buildings. Most building systems are designed to keep the occupants safe, comfortable, and secure while maintaining their privacy. HVAC and IAQ systems in buildings are designed to maintain comfortable and healthy environment inside buildings. Occupants may not necessarily be the direct customers for purchasing, installing, and maintaining the building products in most cases. However, they play a key role in using the products and influencing

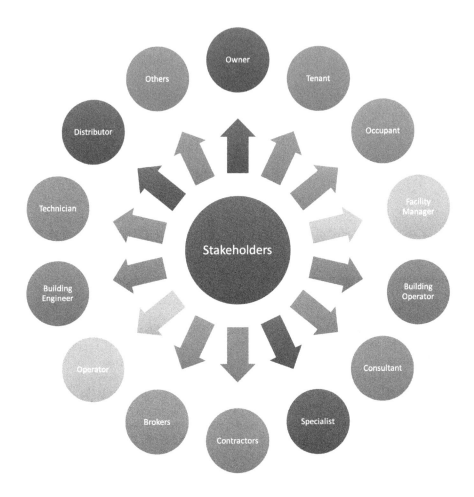

FIGURE 7.1
Types of stakeholders involved in building systems.

the purchasing decisions. In residences, several products (such as smart thermostats, sensors, video cameras, and security systems) are targeted directly to the occupants as they have the power to purchase and maintain the products. In contrast, the occupants play minimum or no direct role in purchasing or maintaining building systems in commercial buildings. In commercial building, some occupants contribute to the operation of building systems, e.g., thermostats and lighting control in private office or small conference rooms, and access control in an R&D laboratory. As emphasized in this section, occupants are the main customers of controls and a control product must satisfy the needs of occupants for it to be successful.

7.1.2 Tenant

Tenant is an individual or an organization renting or leasing a building from the owner. Tenants are responsible for paying the bill to the owner and maintaining the property depending on the contract setup between the parties. It is not necessary that tenant is the owner of a property. According to the census, there are almost 78 million and 44 million owner housing units and rental housing units, respectively [4]. According the Pew Research Center, 37% of household heads rent their homes [5]. Similar to the occupants, tenants are the main consumers of the property. However, they have varying impact on the purchase of systems inside the buildings. Tenants of a leased facility are willing to make enhancements to the property as compared to the people who rent buildings for short period of time. Contracts also impose constraints on possible modifications or functions that the tenant can perform in a building. Occupants are not necessarily the tenants and vice versa although they share many attributes with each other.

Federal government and state agencies also impose constraints on owners or facility management companies to ensure certain level of comfort and safety inside buildings. Another question is "who pays the bill?" Answer to this question determines the targeted customers and the features of a product. A simple example is a homeowner paying the entire utility bill. Another case is the facility management company paying the entire bill in a hotel. It is also possible, yet not common, that a home owner association (HOA) pays the bill (electricity, water, or gas) for the entire community. The HOA divides the bill accordingly among the community buildings.

7.1.3 Owner

As the name suggests, owner is an individual or an organization who owns the building. Many key decisions related to the system in a building are handled by its owner. If owner is not the occupant, the main motive of the owner is to maximize the short-term and long-term gains. Short-term gains include generating continuous revenue from the tenants by keeping the tenants and occupants comfortable and safe while minimizing the investment. If an owner is responsible for paying the energy bills, he/she will be willing to invest on a new equipment that has a reasonable ROI (return on investment), e.g., payback period between 3 and 5 years. Long-term gains and related activities include establishing a brand, delivering the brand value, or increasing the value of the building for future sales. Owners have higher contribution to capital investment inside buildings as compared to the operational expenses.

7.1.4 Facility Manager

Facility manager is in charge of overall operation, maintenance, and upkeep of buildings. He/she performs management tasks and delegates work to other staff members on managing and updating equipment, office systems, and other building systems such as parking, electrical, plumbing, mechanical, retrofits, and change orders. Monetary decisions are normally made by a facility manager after consulting with the owner. It means that facility managers are influential in affecting the owner decisions. Facility managers ensure healthy, safe, and comfortable environment while meeting legal requirements in every condition including emergencies. Facility managers can be directly working for the employer owing the facility or property management company. Facility managers are also involved in formulating a roadmap and plan for the facility to achieve certain performance metrics, e.g., energy and efficiency goals. They also maintain relationships with vendors, owner, staff members, and occupants.

7.1.5   Building Engineer and Building Operator

Building engineer and building operator have very similar roles on the building automation side. Building operator focuses primarily on the operation side of the systems while building engineer oversees the overall functioning of the system including repairs, maintenance, and several design and retrofit aspects. Building engineer has broader knowledge and skill-sets to assess different, diversified situations and propose solutions to solve the problems.

There are many types of operators inside the buildings. As part of the operational side in large commercial buildings, a building operator performs day-to-day functions to ensure proper functioning of building systems that are integrated into the BAS. As the name suggests, they handle several operational aspects such as ensuring equipment functioning, creating alarms, setting up trends for investigating and debugging issues, changing or adding the sequence of operations to improve the working of existing systems, and managing inventory control processes. Building operators are responsible for updating the existing control systems. Understanding their perspective is crucial in the situations when a new control product that replaces the existing control product is launched.

7.1.6   Consultant

Consultants can be involved in a few different stages of a building life-cycle. During the design phase before a building is constructed, consultants create technical specifications according to the owner requirements. Contractors use the specifications to construct a detail design of the building and the systems inside the building. Owners also involve consulting firms—mostly for commercial buildings—during maintenance or operation phase to seek their suggestions on improving the performance and aesthetic of buildings.

7.1.7 Contractor

There are many types of contractors: general, mechanical, civil, and electrical. For building control systems, control contractor, which is part of mechanical contractor, is most relevant to the installation and working of building control systems. Contractor is a broad term that can be used in many ways. In this case, contractor is an individual or organization solely responsible for finishing a specific part of the project. Mechanical contractor is responsible to design, procure, and install the mechanical systems in a building as per design specifications. Mechanical contractor either subcontracts the controls work to a controls contractor or finishes the controls project himself/herself.

7.1.8 Brokers/Representatives

Brokers and representatives connect producers to the contractors and end-customers. Although they are not employed by the manufacturer, they develop strong relationships with the company. They are the face of the manufacturer. They must have good understanding of the entire system (not just technical) to facilitate smooth conversations between manufacturers and contractors.

7.1.9 Technician

The primary function of a technician is to fix the broken component of a system. There are many types of technicians: electrical, mechanical, control, IT, and construction. As part of the facility management team, technicians have refined, narrow focus with higher hands-on experience than others. As they are focused on assembling, repairing, and fixing a specific system or components of the system, they care about the tools that can help them find the problems and fix them quickly and easily. Instead of prioritizing energy efficiency, technicians pay higher attention to the time, parts, and tools needed to fix an equipment or controller.

7.1.10 Utility and Government

Part of the infrastructure including data from meters is owned by the utility companies. Utility companies also provide/implement incentives and rebates to incentive or penalize the end-users. Utilities also deploy demand response programs and sponsor research to evaluate certain technologies that can benefit both the consumers and the power grid at the distribution and sub-distribution levels. Utilities play an important part in advancing the technologies or the adoption of such technologies although they don't have much direct influence on the building owners or the components installed in buildings.

Similar to the utility companies, government agencies (local, state, and federal) provide incentives to the users to take certain actions that are beneficial for the people and the society, e.g., use of solar and wind energy to power the buildings. Safety, health, and privacy issues are the topics of discussion for the government as they consider the holistic picture and the impact of new technologies on the environment and the country. Government allocates a certain amount of budget in researching, marketing, and adoption of technologies used in building systems. Demonstrating the value of such technologies by government agencies can provide a head-start to wide variety of businesses (start-ups, small business owners, large organizations, etc.) in this industry.

7.1.11 Other

There are other diverse stakeholders (internal and external) with differing background involved in building life-cycle management:

- Commissioning Agent: An independent company hired by the owner to ensure that the building systems are installed and working as per design proposal and specifications.

- IT and Security Manager: Similar to the facility manager, IT and security managers are responsible for the information technology and security department of the building, respectively.

- Integrator: In large buildings, there are heterogeneous products from different vendors performing similar functions. Integrator amalgamates the products into one unified product to provide a common interface and enable typical functions from one place. In residences, integrators have been increasing in recent days because several smart products from multiple vendors are found in building systems. Special expertise is needed to make the products work with each other, e.g., alarm.com [3].

- Distributor: An individual or a company who distributes the product from one party to another. Details on the distributor's role are provided later in this chapter.

Other less-affected, but related stakeholders are developers, builders, architects, and construction managers. Figure 7.2 shows the list of key stakeholders and a few items that they consider the most in their role.

7.2 Distribution Channel

Distribution channels involve a number of parties through which the products are delivered to the end-customer. There are three main types of channels: (1) Direct, (2) Indirect, and (3) Hybrid, which are explained next.

- Total cost (design, commissioning, maintenance, and operation
- Brand value
- Short-term value
- Legalities (safety, security, and health)

Owner

- Comfort
- Operational cost
- Short-term value
- Safety and security
- Health and privacy
- Penalties

Tenant and Occupant

- Environment
- Sustainability
- Health
- Safety and privacy
- Technology advancement
- People life

Utility, Government & Others

- Site issues (health, safety, security, privacy, and operation)
- Maintenance cost
- Operational cost
- Code and standard compliance
- Expert dependency
- System usability

Facility Manager & Team

- Design suggestions
- Clear and precise guidelines to contractors
- Specifications reusability
- Meeting budget constraints and owner requirements
- Recommendation on systems with minimal issues

Consultant

FIGURE 7.2
Summary of key stakeholders and their roles/attributes involved in building systems.

7.2.1   Direct Channel

Figure 7.3 shows the flow of product in a direct channel. As clear from the name, producer supplies and distributes the products directly to the customers. Sales made through the websites and stores owned by the manufacturers are considered as part of direct channel sales. In a direct channel, companies enjoy high control on the product quality, sales process, and customer experience. Customers can communicate directly with the producers and vice versa. It means that issues experienced by the customers can be quickly communicated to and resolved by the producers. Discounts on products and new products can be launched faster than other channels.

FIGURE 7.3
Direct marketing channels in building systems.

In this channel, the producer is solely responsible for marketing and generating sales. There are less dependencies on intermediate parties as compared to other channels. Producers are able to collect data from customers and offer better prices to the customers as there are no third-party distributors. One of the biggest drawbacks of direct channel is the capital investment required to open a store. Developing a website is not much expensive, however, continuous update and maintenance of a website while securing customer data against cybersecurity attacks can be time consuming and expensive.

Smart thermostats from Google and Ecobee sold through their websites are noted under direct channel. Other examples of products that are sold via direct channel in buildings are smart lighting switches, smart sensors, demand response adapters, smart garage door openers [7], smart security systems [11], smart blinds [8], smart meters [10], and smart appliances [2]. Although direct channel is prevalent in consumables markets (primarily in residences), it can be found in some parts of commercial buildings, e.g., smart sensors, software tools, equipment firmware update, control algorithm updates, and debugging accessories for commercial building systems are made available by the manufacturers directly.

7.2.2   Indirect Channel

As opposed to a direct channel, indirect channels involve at least one intermediate party that interfaces with the end-customer and sells them the products. Figure 7.4 shows a high-level schematic of the physical flow of a product in an indirect channel. The number of intermediate parties and their role are dependent on the types of product, business, and the customer. Indirect channels are most prevalent in commercial buildings.

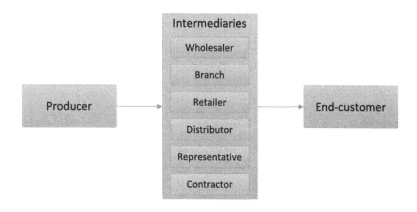

FIGURE 7.4
High-level schematic of indirect channels in building systems.

Indirect channels offer several benefits to the producers. Producers can leverage the experience, sales force, and setup of intermediaries to sell high quantity of products to their customers. Producers do not necessarily need to setup a large infrastructure to manage, advertise, and sell their products. New producers can quickly enter the chain of selling products through the intermediaries. Producers may not need to worry about selling products to the end-customers as the producers can make an agreement with the intermediaries. Selling products through intermediaries may provide a cushion to the producers in case of unexpected events or short-term crises. If the producers sell a large quantity of products to wholesalers or retailers and any of the retailers/wholesalers go out of business, the revenue of producers can be significantly impacted. Because a portion of the responsibilities is shifted to the intermediaries, producers are held less liable in case of distribution problems or conflicting situations.

As compared to a direct channel, the price of a product is higher and the profit margin for the producer is lower in an indirect channel because of the expenses incurred at the intermediaries during multiple stages. Producers also lose touch with the final customers. Loss of information is possible during communications from producers to end-customers and vice versa. The quality of service and product is highly dependent on the entire chain. Any weak link in the chain can impact the end-customer. Therefore, producer, end-customer,

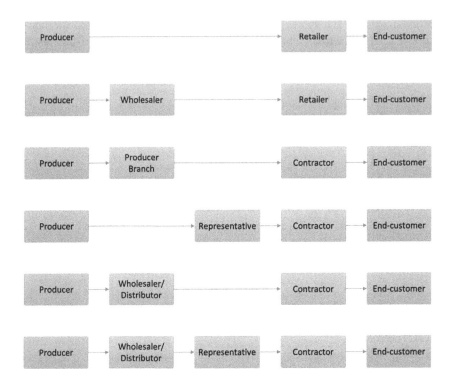

FIGURE 7.5
Indirect distribution channels in building systems.

or an intermediary has less control in an indirect channel as the power is distributed across the board.

Figure 7.5 illustrates several forms of indirect channels in building systems. The simplest form of indirect channel involves only retailer as an intermediary. This type of channel is commonly found in consumables market. Smart thermostats, sensors, and devices that are sold through retail stores such as BestBuy, HomeDepot use an indirect channel. Depending on the quantity/price of products and producer/retailer requirements, wholesalers can also be involved in procuring products and selling them to the retailers.

In commercial buildings, a few large control producers have setup regional offices, called "branches," to establish closer relationship with the final customers. Branches sell their products to the local contracting firms or licensed contractors who install the products inside the buildings. If the producers perform the job of a contracting firm, the entire flow is considered under direct channel. Instead of a regional office owned by a branch, the products are distributed to a contractor through a wholesaler or representative. The main difference between a wholesaler and representative is that wholesalers often store bulk products in their warehouses while representatives neither own a warehouse nor store any products. Instead, representatives are licensed agents or brokers trusted by the producers to place and ship orders to the contractors.

The most complex form of an indirect channel is a 3-level channel consisting of wholesalers/distributors, representatives, and contractors; shown at the bottom of Figure 7.5. In this channel, wholesalers obtain products from producers. The role of producers is minimal afterward. Representatives or brokers transfer small quantity of products to the contractors who install or deliver the final product to the end-customers.

7.2.3 Hybrid Channel

Hybrid channels use both direct and indirect channels to perform certain types of tasks and activities. Hybrid channels overcome the disadvantages of both direct and indirect channels. For instance, direct channels provide direct communications with the end-customers while indirect channels leverage the infrastructure set up by the intermediaries. Types of channels are decided based on business segments, customer type, and other criteria that are critical and relevant to the product. For instance, consumable products are launched through a hybrid channel, e.g., Google Nest [1]. Figure 7.6 summarizes the types of direct and indirect distribution channels exist in the building systems.

As an example of hybrid channel, Google Nest offers both direct and indirect channels to their customers. Figure 7.7 shows the channel partners, facilitators, and service outputs associated with the product. The product offers spatial convenience as the customers can easily buy the products with fast delivery options. Customers have options to purchase the product directly through the Google website, Internet retailers (Amazon, eBay, etc.), and store retailers such as BestBuy and HomeDepot. The product is available in variety of colors and configurations to accommodate the customer needs. Service add-ons are available to the customers, i.e., customers can install the thermostat by themselves or through a contractor. Facilitators, who are not the primary intermediaries but help in the distribution process, are transportation companies, storage buildings, and banks.

7.3 Channel Power and Conflicts

Producers can influence the behavior of channel intermediaries although producers may not be directly involved throughout the process. Channel

FIGURE 7.6
Marketing channels in building systems.

power refers to the ability of producer to affect the behavior. Channel power of a producer depends on the company's brand, supply and demand of the company's products, intellectual property, the company expertise, and the financial/non-financial incentives supplied to the channel partners.

Choosing a right channel(s) is quite complicated as each channel has its pros and cons as described in the previous section. Selection of channel varies according to geographical locations and cultures. In case of indirect channels, maintaining integrity and quality of the channel partners requires continuous interactions between producers and intermediaries. It is also important to not

Downstream Partners	Service Ouputs	Facilitators
Fast delivery (via direct and retail channels)	Store retailers (BestBuy, HomeDepot, etc.)	"Working with Nest" partners
Variety (multiple colors and configurations)	Internet retailers (Amazon)	Software companies on Developer Platform
Spatial convenience	Independent mechanical contractors	Transportation (Fedex and UPS)
Service addons–multiple installation options and backup		Others (advertising firms and banks)

FIGURE 7.7

Channel partners, facilitators, and service outputs of Google Nest.

only choose an appropriate channel but also make necessary modifications based on new innovations, competition, and changing customer needs.

Channel conflicts can occur at multiple levels between the intermediaries, consumers, producers, and other channel partners. A few reasons for these conflicts are lack of proper setup and agreement between producers and intermediaries, clashing of sale territories, offering multiple competing channels for the same customers, and differences in perception, goals, and services. Usually, many building companies offer multiple channels for consumable products. Multiple channels competing to attract the same customer can result in channel conflicts. The following example illustrates the channel power and associated conflicts when multiple channels are selected.

Example on channel power and conflicts: Google Nest [1] retail partners offer spatial convenience to its customers. As the product focuses on aesthetics and technology-enabled features, it is important for the customers to see and feel the product. Furthermore, because of the product's emphasis on easy installation, the company wants to attract customers that are interested in avoiding installation/service fee, similar to "Do It Yourself" campaign of Home Depot [9]. Initially, end-customers are directly targeted and middleman (i.e., contractors and technicians) is removed from the picture to increase the margin. Google sells directly through their websites and its downstream partners—store retailers (BestBuy, HomeDepot, etc.), Internet retailers (Amazon), and independent mechanical contractors.

Recently, Amazon has stopped selling the Nest thermostats and other Nest products because of company's future directions and evolution of products that are competing directly with each other. As a response, Google pulled off YouTube from Amazon devices because of privacy violations. Due to the size of the companies and growing competition between them, there are continuing conflicts between the companies [6].

Nest does not have much coercive power because of its limited market share. Instead, Google holds a combination of legitimate and referent powers because of the company' reputation over the past several years. In contrast, Apple has high channel power because of the company's expertise (knowledge, patents, and technology), high demand, high market share, and well-reputed brand.

Key Takeaways: A Few Points to Remember

1. Buildings have a complex product distribution structure with involvement of several stakeholders at multiple levels.

2. A few users or stakeholders in buildings are owner, tenant, occupant, facility manager, building operator, consultant, specialist, contractor, broker, building engineer, technician, distributor, integrator, IT manager, utility, and government agencies.

3. Control product can be targeted for a single stakeholder or a class of users as it is very challenging to target all the stakeholders in the value chain. At the same time, the product should not affect the other stakeholders in a negative fashion.

4. Understanding the role and involvement of stakeholders and the entire delivery chain is important in the product roadmap and the features.

5. Direct, indirect, and hybrid channels are three types of distribution channels found in buildings.

6. With varying number of intermediaries in building distribution channels, the complexity increases with the increase in the number of intermediaries.

7. Channel power and potential conflicts need to be evaluated when introducing a new or multiple distribution channels to a customer.

8. Common attributes of a channel are responsiveness, communication level with customers and channel partners, complexity, initial capital investment, infrastructure setup, legal factors, and the level of responsibilities.

9. Developing a product (or the product features) and the selection of channel go hand-in-hand for successful delivery of product.

Bibliography

[1] Google Nest. `https://nest.com/`. Accessed: 2020-03-22.

[2] AJ Madison Inc. 7 Wi-Fi Enabled Smart Appliances for a Smarter Home. `https://www.ajmadison.com/learn/7-wi-fi-enabled-smart-appliances-for-a-smarter-home/`, 2019. Accessed: 2020-11-29.

[3] Alarm.com. Benefit-Cost Ratio (BCR). `https://www.alarm.com`, 2020. Accessed: 2020-03-22.

[4] Robert Callis, Patricia Holley, and Daniel Truver. Quarterly residential vacancies and homeownership, second quarter 2020 (Report No. CB20-107). *Retrieved from the United States Census Bureau website: https://www.census.gov/housing/hvs/files/currenthvspress.pdf*, 2020.

[5] Anthony Cilluffo, Abigail Geiger, and Richard Fry. More US households are renting than at any point in 50 years. *Pew Research Center*, 2017.

[6] Steve Kovach. Amazon will stop selling Nest smart home devices, escalating its war with Google, Mar 2018.

[7] Molly Price. Best smart garage door controllers of 2020: Chamberlain MyQ, Tailwind and more. `https://www.cnet.com/news/best-smart-garage-door-opener-controller-of-2020-chamberlain-alexa-google-homekit-myq-tailwind-nexx/`, Nov 2020. Accessed: 2020-11-20.

[8] Luis E. Chavez Saenz. What are smart blinds? `https://www.lifewire.com/what-are-smart-blinds-4685778`, Feb 2020. Accessed: 2020-11-29.

[9] Zeynep Ton and Catherine Ross. The Home Depot, Inc. *Harvard Business Review*, 03 2008.

[10] US Department of Energy. Advanced metering infrastructure and customer systems: Results from the smart grid investment grant program. Technical report, US Department of Energy, September 2016.

[11] Megan Wollerton. Best DIY home security systems of 2020. `https://www.cnet.com/news/best-diy-home-security-systems-of-2020/`, Nov 2020. Accessed: 2020-11-20.

8

Product Marketing Management

> "The greatest strategy is doomed
> if it's implemented badly."
>
> — Bernhard Riemann

Marketing management is an act of identifying and satisfying the needs of customers through products, services, experiences, and information. Marketing is used in a broad sense in this chapter. Marketing includes creating, communicating, and delivering value to the customers. Marketing management is one of the most crucial steps for any successful company because the marketing activities involve appropriate product selection (along with product features from customers perspective), market targets and segmentation, product strategy, competitive/growth strategy, and strategy to establish and strengthen customer relationships. Definition of a company (or a set of products offered by the company) from marketing sense is broader than the definition of products. Figure 8.1 expands a few definitions to their corresponding broader definitions from marketing perspective.

8.1 Product Research and Segmentation

8.1.1 Product Market Research

Product market research consists of three main tasks: (1) state-of-art investigation, (2) generating ideas based on customer needs, and (3) evaluating market growth. State-of-art investigation is a step to identify the latest technologies of multiple components or systems used in buildings. This investigation also includes academic research or R&D work, which has the potential to be demonstrated and implemented in products, i.e., technologies at high TRL (technology readiness levels [26]). Technologies and methodologies in modeling, sensing, software, control, communication, and testing areas need to be scouted continuously as discussed in Chapters 1, 2, 3, 4, and 5.

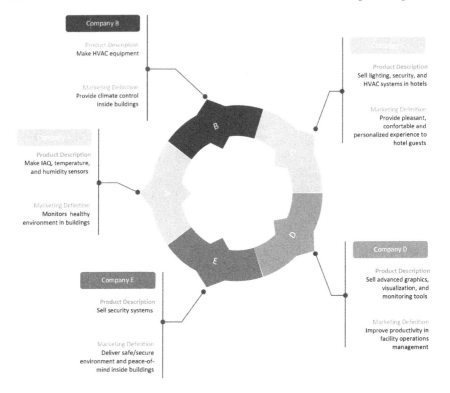

FIGURE 8.1
Product description and corresponding market definition of hypothetical companies.

Ideas that solve customer problems can be acquired by direct or indirect interactions with end-customers. Customers are most likely to discuss the problems with the intermediaries closest to them. In non-consumable markets, the distribution and marketing channels are mostly indirect and thus the end-customers may not feel comfortable in sharing every issue or potential new product features with the producers in detail. In these scenarios, producers rely on the intermediaries to gather and provide feedback from the end-customers. In the buildings sector, most of the business is done through B2B partners. Therefore, it is essential to not only solve end-customer problems but also pay attention to the intermediaries' problems.

A few options on obtaining new ideas are: brainstorming sessions with trusted group of end-customers and stakeholders, informal sessions between technical staff members, surveying end-customers and stakeholders, involvement of stakeholders in product planning, real-time feedback on social media, conferences, and trade shows. Besides generating ideas from end-customers and intermediaries, the ideas can be sourced from internal employees, especially on internal processes, efficiency, product quality and cost. In most situations, end-customers or other intermediaries can provide suggestions only toward incremental product improvements due to many reasons. If developing a completely new product, it is recommended to use multiple approaches rather than solely relying on customers suggestions for validating the product needs.

Usually, customer need is a good indicator of the market potential. Without customer needs, there is no market. If there is no existing market (or future potential) for a certain product, it is obvious for a company to divest (not invest) its resources on a product targeting that market. Quantification of potential market is absolutely needed for both existing products and new business ideas or products. Currently, building systems (excluding construction) industry has huge potential with more than $150 billion market in the US [30, 31, 19, 17, 20, 27, 21, 13, 34, 33, 33]. Figure 8.2 shows the current market potential [30, 31, 19, 17, 20, 27, 21, 13, 34, 33, 33] classified into a few categories.

8.1.2 Customer Segmentation and Targeting

As the customer needs are quite varying and different from each other, it is typical for a company to develop several products that connect with a diverse range of customers. Customer segmentation is about categorizing a large group of customers into small buckets so that the company can better understand and satisfy their needs and wants. This way the company can effectively connect with the customer and communicate the value of its product and brand. Market segmentation also helps the company in deciding the scale of targeted market and how the market size aligns with the company strategy/size. For instance, as compared to a start-up, a large company can relatively easily target large market segments because of their existing customer base in those markets and investment resources. A list of possible criteria to categorize the customer market in buildings is shown in Figure 8.3. Details on the segmentation options are furnished next.

1. Customer: Distribution and marketing channels in buildings have several intermediaries and stakeholders. Although end-customer is mostly prioritized for any product or service, intermediaries and stakeholders are intentionally targeted for specific products because the intermediaries and other stakeholders: (1) influence the decision of end-customers; (2) contribute to the overall working of the building ecosystem including its distribution channel; and (3) communicate and deliver the value of product and brand in many cases. Stakeholders and intermediaries are explained in Sections 7.1 and 7.2, respectively.

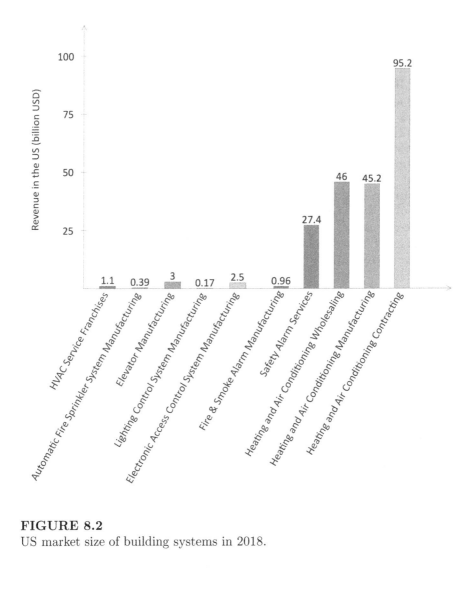

FIGURE 8.2
US market size of building systems in 2018.

2. Building Type: Principal activity in a building decides the type
    and usage of the building. Product needs and marketing strategy
    are highly dependent on the type of building. As an example,
    control and quality requirements of a hospital will be much higher
    and different than that of a university building. Control product
    developed for a clean laboratory must offer precise and accurate
    control of inputs and outputs. Office, warehouse, storage, food,

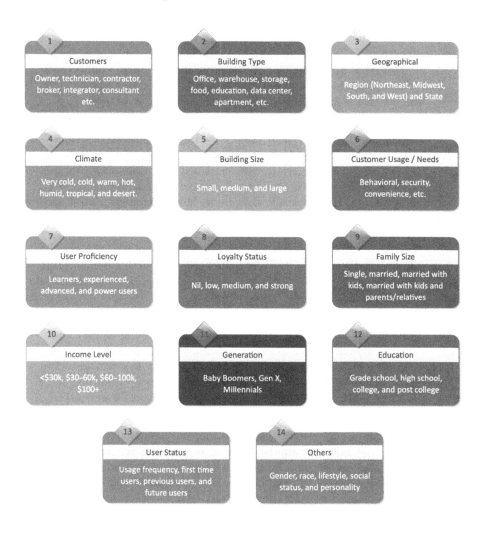

FIGURE 8.3
Market segmentation options in buildings.

education, data center, apartments, and laboratory are a few building types; refer to Section 1.2 on buildings classification.

3. Geographical: Geographical segmentation can be concluded by region, states, and countries. According to CBECS [32, Table B-3] geographical regions are Northeast, Midwest, South, and West. Regulatory policies, incentive mechanisms, buildings standards, building codes, and guidelines vary from one state to another.

Therefore, clear understanding of such factors leads to not only better adoption of products but also better differentiation of the products from the competitors' products. It is vital to consider the factors that are dependent on the geography, and more importantly understand the differences between the regions, states and countries to decide product and corresponding marketing strategies.

4. Climate: There are many ways to categorize climate into zones. According to the article [28], US climate consists of tropical, dry, moist subtropical, moist continental, polar, and highlands as major zones, which are further classified into sub-zones. Another classification, which is based on the usage and functioning of HVAC system, is very cold, cold, warm, hot, humid, tropical, and desert. Climate classification enables specific design and use of technologies in a product. In very cold environments, the control product needs to provide a defrosting mode and the equipment/controller components need to sustain and work well in low-temperature conditions. New certifications may be added to the list in these situations.

5. Building Size: Small, medium, and large are three available sizes on buildings. Selection of products in a building is influenced by the building size. Small commercial buildings may have less complex building system and usually demand less features than large commercial buildings. A small mom-pop restaurant may have higher monetary constraints and lower flexibility as compared to a big-box store such as Walmart.

6. Customer Usage/Needs: Dividing the market based on customer wants and needs is another way of segmenting the market. The wants and needs could be derived from the usage and preferences of end-customers. The requirements can also be developed by offering a product that combines the features of existing, multiple products because a certain number of end-customers use only some functionality of those products. An end-customer may need high security product with remote accessibility regardless of size and type of buildings because the building is located in a high-crime area.

7. User Proficiency: User proficiency—learners, experienced, advanced, and power users (also called experts)—determines the level of complexity that can be embedded into a product. If the targeted market constitutes learners and low experienced users, it is recommended to reduce the level of advanced features with steep learning curve because the learners and low experienced users have to learn not only the existing system but also the advanced features in short period of time. This may create confusion and wrong initial perception about the product and the brand.

Programmable thermostats faced many challenges in terms of correct usability because of their complexity and low user proficiency in homes. A study [14] shows that programmable thermostats are not only difficult for users to understand but also difficult to program. This led to ineffective use of the device, causing higher energy consumption on average than the systems with simple, non-programmable thermostats. As a result, the U.S. Environmental Protection Agency (EPA) suspended the ENERGY STAR certification for programmable thermostats in 2009.

8. Loyalty Status: Loyalty status includes customers with no, low, medium, and strong loyalty toward a company brand. It is difficult to shift strong-loyalty customers from one brand to another as compared to the customers with low or medium loyalty. Loyalty is mostly applicable to consumable markets in this industry. In B2B market, certain types of users find it difficult to switch the brands because of steep learning curve involved in the process. Each company offers a variety of software and hardware tools that may not be interoperable with each other. Therefore, changing a component from one manufacturer to another requires significant amount of effort and learning.

9. Family Size: Family size is only valid for consumable markets, primarily in single and multi-family homes. Single, married, married with kids, married with kids and parents/relatives are a few ways to segment the market in this category.

10. Income Level: Income level molds the preference of customers. This category is reasonable to only consumable markets. Energy bill of low-income households is up to 20% of their annual income. In contrast, for the high-income households, the energy bill is less than 2% of their annual income [29]. Therefore, energy and monetary savings are high in the priority list of low-income customers while convenience, aesthetics, and other features are preferred by high-end customers.

11. Generation and Education: Stakeholders with different generations are attracted to different aspects of a product and company. Baby boomers and highly educated customers are wealthy with high spending power. Highly educated customers are more analytical, paying larger attention to details while understanding the consequences of the actions. On the other hand, millennials are highly tech savvy, conscious to the environment, and socially aware. Although millennials may not have high spending power, they have another 40–50 years of life expectancy. Engaging them in early stage can be beneficial in long run.

12. User Status: In this category, the users are profiled based on their usage of systems every day. Large buildings with BAS

have higher usage rates as opposed to small-office type buildings because large buildings have a dedicated team to maintain and operate the buildings. Buildings with large usage of systems are comfortable spending significant amount of resources. Status of users are divided as per their awareness and usage. Along with the first-time users, previous users, repeat users, and future users, it also includes the stakeholders and end-customers who are aware of the company/product but have never tried the product.

13. Others: A few other attributes to segment market are gender, race, lifestyle, social status, and personality, culture, and behavior. These attributes are not prevalent in the building systems, but applicable to specific small market groups. For instance, single, progressive users with high standard of living (e.g., a celebrity) demands a product that is environmentally/sociable responsible with emphasis on aesthetics, functionality, and risks. Such segmentation helps the company decide if the product will provide a cookie-cutter solution or a specialized custom solution for a certain class of users.

Large companies in this industry are primarily focused on B2B (e.g., engineering firms, and mechanical contracting firms) markets with increasing interest in B2C (consumers/end-users) markets in North America. In the US, companies gather the needs and wants from a variety of customers at multiple levels. Along with the market research methods mentioned earlier in this section, marketers utilize experience from people who have worked on the site/field. It means that these people have either worked at the B2B firms or interacted closely with both B2B and B2C partners.

8.1.3 Market Adoption Behavior

Market adoption curve is a way to classify the market based on the behavior of users to adopt a new technology. Figure 8.4 shows an adoption curve with five major types of users: (1) Innovators, (2) Early adopters, (3) Early majority, (4) Late majority, and (5) Laggards. This classification of users identifies their response to innovation during different penetration stages.

Innovators are high-risk takers as they are ready to adopt innovative products and services that have not been launched in the market before. Although innovators are small in number, it is necessary to gain the trust of innovators to proceed to the next stage. Innovators also act as marketers for the next types of users, i.e., early adopters. Early adopters are the second fastest adopters of new technologies. Innovators and early adopters are generally younger in age and they have cushion to absorb the failure of products, i.e., their income level is not in the lowest category.

Early majority customers are medium risk takers with belief in innovation. Product enters the growth phase when early majority customers start using

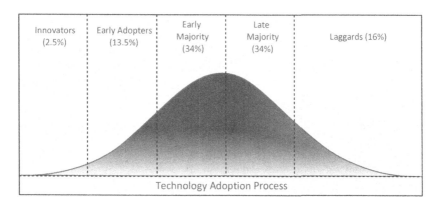

FIGURE 8.4
Market adoption curve.

the product. Late majority customers followers and averse to the risks. They rely on adopters and early majority customers to validate the technology. They also wait for the price to be reduced and the product quality to be improved. Laggards are resistant to any change. They are last in the chain switching to the technologies, especially when no other option is available to them. For example, laggards transitioned to smart phones from a traditional keypad phones because major cellular/communication companies stopped selling the traditional keypad phones.

8.2   Competitive Analysis and Product Positioning

Market segmentation decides the customer type, customer needs, and corresponding market size. Competitive analysis evaluates the products and marketing strategies of competitors. SWOT analysis is a technique to assess the strengths, weaknesses, opportunities, and threats of a company, business area, or a product. Figure 8.5 shows a schematic of SWOT analysis with distinct characteristics in these four sectors.

Using SWOT analysis, a company can organize the key attributes to decide the organizational or product strategy. With changing environments and growing competitions, it is essential to evaluate the strengths, weaknesses, opportunities, and threats continuously at periodic intervals. Strengths and weaknesses are internal to an organization, i.e., the company can modify them by taking internal actions. It is possible that there are certain areas that swing between strengths and weaknesses, especially at the competitive

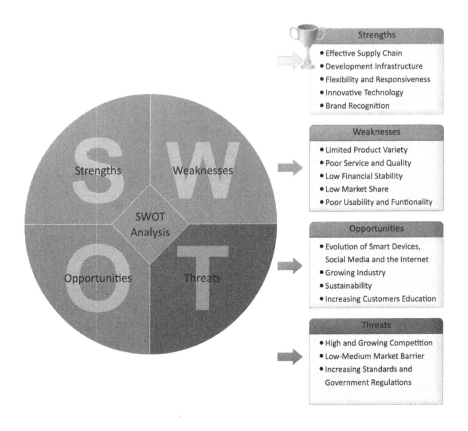

FIGURE 8.5

SWOT (strengths, weaknesses, opportunities, and threats) analysis.

products. Instead of catching up with the competition continuously, the company should look at the SWOT analysis and decide the areas that needed to be strengthened and aligned well with the business strategy.

Conversely, opportunities and threats are external to a company, i.e., a company cannot necessarily change them but the company can utilize them for its advantage. There are several threats to the companies that can affect their business and future growth. A few of them, in addition to the ones described in Figure 8.5 are channel dependency, global climate change, regulatory standards, volatility in the commodity prices, and recession/inflation.

At the same time, there are tremendous opportunities because of change in technologies, availability of more efficient processes/systems, and increased awareness in customers. Most of these technological advancements are fundamental and they could be applied to all companies from product development to supply chain. For example, AI can be applied to not only the products to reduce energy consumption but also predict the market demand so that the manufacturing plants can prepare accordingly in advance. It is critical to make sure that the corporate, marketing, supply chain, and IT strategies are well aligned and linked to each other. For example, since AI is a new cutting-edge technology for building systems, AI in any product can be very difficult (not technically) to implement and execute if the corporate strategy does not allow innovation and automation to solve customer problems. Another example is to expand the portfolio into cloud computing for buildings. If part of this idea is not supported by the company's corporate strategy, does it mean that the company needs to change its corporate and business strategies? SWOT analysis is deemed as a very important intermediate step during the entire evaluation process. It is not just one-time analysis, instead the SWOT analysis needs to be checked and updated continuously to ensure that (1) company is heading in the direction to grow the business, and (2) there is close alignment between business, marketing, IT, and supply chain strategies.

8.2.1   Positioning and Competitive Advantage

Marketing strategy includes positioning of a company (or a product) in the minds of customers and differentiating the company from its competitors to drive growth. The first step is to identify the list of basic attributes that the customers expect from a company in the buildings industry. The next step is to determine the attributes that are offered by almost every major brand in the industry, e.g., many brands target to provide reliable, stable, and responsible solutions with a variety of options. Another activity in this process includes obtaining customers ratings on products from different companies. By sorting the ratings from customers, a company can better understand the marketing strategies and behavior of competitors, e.g., is the company changing its strategies toward innovative new technologies or is the company making incremental changes to stay in the business that was established a while ago but is no longer the core business. After understanding the attributes along with the SWOT analysis, the last step is to develop a differentiation strategy for a product or the company by leveraging the current positioning.

Differentiation is one of the vital attributes to success. If there is no differentiation, there is no long-term sustaining growth. Companies release new products with enhanced features to distinguish themselves from others. Differentiation does not have to be technical; it may include service, quality or other attributes mentioned as part of SWOT analysis in Section 8.2. The B2B customers in the commercial market expect an excellent customer support and long-lasting, quality products, and thus are willing to pay higher prices than the competitors.

A few major control companies differentiate themselves on technical features, performance quality, reliability, and customization. These companies offer several products with varying features based on customer needs. They may offer very high quality and reliable products because of detailed internal testing and sophisticated engineering practices. Their products last for 20–30 years in many cases despite their expected life of 10–15 years. As there are variety of customers with individual needs in the same market (e.g., one hospital can have completely different requirements from another hospital down the street depending on the stakeholders' progressiveness and priorities), the companies offer solutions with full customization capabilities although there is a steep learning curve in using and modifying the solutions as per individual needs.

Some companies differentiate on excellent customer service, e.g., providing technical support for 30-year-old products, helping customers in fixing issues even after the warranty had expired. As compared to traditional HVAC systems, VRF companies differentiate themselves on energy efficiency and lower installation costs. Figure 8.6 shows an example of visual competitive analysis of the products offered by three hypothetical companies: A, B, and C.

Figure 8.6 shows that Company A, Company B, and Company C sell two, three, and two products, respectively, in terms of cost, processing speed, display size, energy savings, and usability. Company A offers expensive, modern products that save energy with high processing speeds. These products are aimed toward high-end, progressive market. Therefore, the products of Company A are mostly in top right quadrants. Company B offers products that are mediocre in every aspect, i.e., they are aiming for affordable markets while keeping some features. Company C has diverse portfolio selling products for two completely different markets, selling to the people (a) who worry about the price the most and may not care about aesthetics or features, (b) who want to purchase their products with several features but they are not ready to pay premium price yet, and (c) who are the existing customers of Company B but looking to upgrade their product.

Such visualization tool helps not only in deciphering the strategies and direction of other companies but also finding a white space that is not well targeted by the existing products. It is clear from Figure 8.6 that there are many open slots that can be considered during detailed competitive analysis. It is possible that some open slots may not be technically feasible or there is not sufficient demand in those areas. For example, creating a device with large display size and high processing speed cannot be developed with the lowest list price because of expensive components. Although the visual diagrams help substantially in portraying the holistic picture of companies' portfolio, comparing individual products on many other features using similar graphics can be cumbersome and overwhelming. Therefore, a tabular form to compare the details of specific products is useful as shown in Table 8.1.

Table 8.1 compares the features of a company's thermostat against the features of five different thermostats. The table is just an example, which

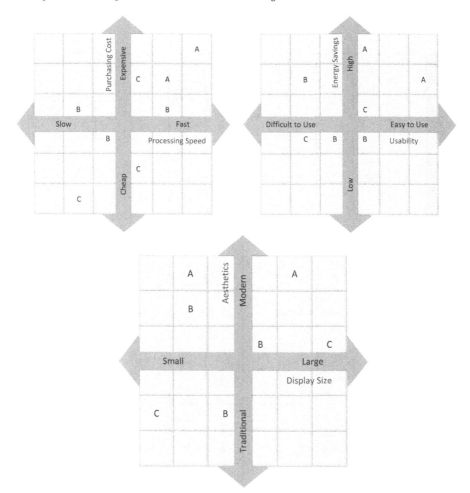

FIGURE 8.6

A visual competitive analysis of products offered by three hypothetical companies.

can be extended or modified to satisfy individual marketer needs. In this example, all the products are fictitious. Products can be chosen from a single company or a group of companies depending on the goal. The cost, screen size, sensors, processing power, remote accessibility, touchscreen display, and communication protocol features of the product are summarized in the table. This type of analysis is performed to jot down the detailed requirements of the next product that your company is planning to target. It also helps in developing a pricing strategy of the product.

Company Products	Product X (Your Company)	Product A	Product B	Product C	Product D	Product E
Cost (in USD)	$149	$199	$199	$99	$150	$249
Remote Access	✓	✓	✗	✓	✗	✗
Screen Size	2.5in	3.5in	2.0in	2.5in	4.2in	4.5in
Built-in Temperature Sensor	✓	✓	✓	✓	✓	✓
Built-in Humidity Sensor	✗	✓	✓	✓	✗	✓
Other Sensors	Occupancy	Occupancy	Occupancy, Light	None	Occupancy	Occupancy, Light
WiFi	✓	✓	✓	✓	✗	✓
Touchscreen Display	✓	✗	✓	✗	✗	✓
Protocol	BACnet	BACnet, Modbus	Proprietary	Proprietary	Modbus	Proprietary
Computational Power	Medium	Medium	High	Low	Low	High

TABLE 8.1

A fictitious competitive chart of residential control products.

8.2.2 Monitoring Competition

Monitoring the behavior, strategies, and products of both existing and new competitors (including direct and indirect) is an input to the marketing strategy. By analyzing the competitors' activity at regular intervals, the company ensures that its brand and product strategies are applicable and up-to-date while maintaining sufficient product differentiation and competitive advantages. If a company started offering Wi-Fi connectivity a few years back, the Wi-Fi feature differentiates the product from its competitors because Wi-Fi was not prevalent in the controllers a few years back. However, many companies, especially in residential market today, are offering control products with Wi-Fi connectivity. Therefore, Wi-Fi is no longer considered as point of parity. Continuous efforts from a company are needed to understand the positioning of the competitors, which encompass their strategy, market share, growth rate, customer perception about their brand/products, and their threats to the company.

8.2.3 Growth Strategy

In terms of driving growth, there are many ways to drive growth. Common growth strategies are:

1. Sell more frequently to the established customers by offering new products, e.g., in 2019, Mitsubishi launched Connect+ [12], which is a software and hardware solution that monitors and controls the HVAC equipment in buildings.

2. Upgrade and sell to the existing customers by adding additional value to the current products.

3. Mergers and acquisitions to expand in related markets; e.g., Johnson Controls merger with Tyco provides holistic solutions in a building such as fire, safety, security, and air-conditioning [22].

4. Expansion in international markets, e.g., joint venture of Carrier with Toshiba [18].

5. Expanding the existing market by building reputation and brand value.

8.3 Product Strategy

Gaps in market segmentation are identified using the products' competition chart described in Section 8.2.1. It is possible that many products can be developed because of several white spaces in the market. In that case, how a company should approach the problem and what product(s) should be

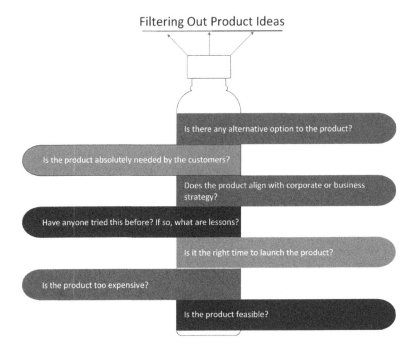

FIGURE 8.7
Ways to filter out the product ideas.

developed? In fact, there may be several ideas to develop the same product. Figure 8.7 shows a few questions that can be asked to filter out the ideas and choose the best idea. A good marketer or an organization would prefer to reject a bad idea in earlier stages of the product development process to minimize the wastage of resources.

8.3.1   Product Selection and Introduction

Product strategy determines if a company is interested in pursuing a completely new product or adding new features to the existing products. In both cases, the company needs to decide—as a crucial part of a marketing plan—if the product needs to be (1) fully developed in house, (2) re-branded only, (3) purchased from a company and re-branded with value addition, or (4) partly developed with commercially available components. These decisions involve trade-offs between control, quality, time, value, and cost. If a company is well equipped with resources (facilities, people, infrastructure, and skill sets) to develop a product in house, the company can better control the timeline and quality of the product. If the company is new in certain area lacking some of the resources, re-branding is a better solution for quicker results. During re-branding, the company has much less responsibilities from design, manufacturing, and supply-chain perspectives. Brand awareness is another factor that needs to be considered while making decisions on product strategy.

8.3.2 Project Management

There are two common approaches adopted while managing a project: (1) stage-gate process and (2) time-bound agile process. Stage gate process, a variant of the product development decision process [24, Chapter 15], uses a waterfall approach in which the entire project is divided into sequential activities. The activities go through different stages/gates where the decisions are made on the next steps. For example, multiple project ideas will get through the screening gate to be selected for the development. The product concept can be determined by analyzing the competitors' products and the needs of customers. Feedback from customers, trade-shows, internal prototypes, and suggestions from experiences employees who have directly worked with customers in the past can be used to solidify the concept. This type of methodology is typically selected for POCs, prototypes, research, and uncertain/long projects, especially related to hardware.

Time-bound process is an agile process to manage a project. In this process, the release date of the product is decided ahead of time. The number of features is compromised in case of delays or uncertain events. As opposed to stage-gate process, activities here are conducted in parallel with rapid prototyping, development, and testing. There are frequent interactions with stakeholders with quick, multiple iterations. In short, there are many mini projects going in parallel. This methodology is very commonly used in software projects.

8.3.3 Pricing Strategies

Pricing strategies comprise setting up the initial price of a product and approach to modify the price of product during later stages based on the market demand, sales, competition, and other factors. From pricing perspective, most products are divided into two major categories: (1) incremental product with enhanced version and features, and (2) completely new product. For the incremental products, the goal is to have the existing customer upgrade and potentially sell to a few more customers. Using historical sales data and market share, a company can estimate the number of customers that are likely to upgrade. Based on the estimates the price is determined, which is typically slightly higher or the same (if competition is higher) as compared to the existing product. It is usually the features of the products that get compromised in case of insufficient profit margins.

For a new product, a 6-step process [24, Chapter 16] can be followed for large projects. In this process, the target audience and estimated demand are determined ahead of time. The next task is to estimate the cost by figuring out the resources (people, equipment, facility, manufacturing, etc.) needed to finish the project. Based on the prices of existing products of the company and other competitors, product/sale managers estimate the maximum amount that the customer will be willing to pay. Markup pricing strategy is used to cover overhead, research, and other costs. Break-even chart may also be used to determine the minimum targeted sales volume. Although many companies mostly charge their customers for hardware or equipment while providing software for free. However, this might be changing in the nearby future as many companies have started to transition toward digital technologies. For specific types of equipment, a discounting or sale policy is implemented.

8.3.4   Advertising

Advertising in general achieves the following functions:

- Introduces the company and products

- Explains products including their features

- Shows benefits and usage of product

- Delivers the value of product and brand

- Reminds and educates the customers

- Makes the customer aware about promotions and incentives

There are many ways to advertise and promote products in the industry: social media, TV, Internet, mobile and desktop applications, cinema, radio, emails, pamphlets, newspaper, and magazines. Most of the advertising methods are applicable to consumable markets, but some methods can be used to target B2B partners, e.g., trade shows, conferences, specific newsletters and magazines such as ASHRAE [4], AHR Expo [1], RESNET [10]. Digital advertising using the Internet and social media is prevalent and effective to reach out to specific customer types while staying within the budget. Advertiser should be informed about the awareness-level of brand or product in customer's mind before creating a marketing strategy. The lowest level of awareness corresponds to situation when a customer has heard about the brand or product, but he/she does not understand the details. The maximum awareness-level reflects when the customer has tried the product and used most of its features. In this stage, the customer is ready to conclude his/her satisfaction or dissatisfaction with the product.

8.3.5   Product Stages

Product life cycle consists of four stages as shown in Figure 8.8:

1. Introduction: This is the first stage in which sales are very small as the product is recently released to the market. The profits are negligible, the product may be in loss because of new product expenditures. Marketing and advertising investments are high which increase the customer awareness-level. People are testing the product and its concept.

2. Growth: During this stage, sales are increasing at a much higher rate as the product is validated and majority of customers start using the product. Profits and awareness-level of customers are high in this phase.

3. Maturity: In this phase, sales are almost stable with very high awareness-level. The company may continue to promote its product or add new product features to expand the targeted market and maintain the revenue. Profits are constant or declining during this stage because of increasing competition and marketing investments. Product prices may be stable or reducing because of high competition.

4. Decline: Customers begin to lose their interest in the product during this stage. As a result, sales are declining and the profits are decreasing. Cost increases because of reduced number of sales and economies of scale.

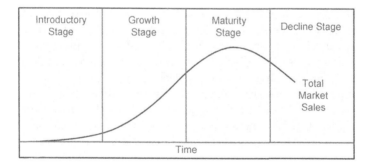

FIGURE 8.8
Product revenue as a function of time and stages.

Example on product stages: Apple HomePod [8], a product in the smart home category to help users on everyday tasks, was launched in June 2018 at WWDC [3] with high focus on the quality of music provided by the speakers at home. User convenience and easy integration features (e.g., Siri integration, vast music library, automatic connection, and spatial awareness) are also the

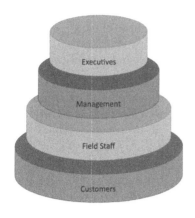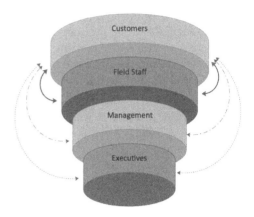

Traditional Organization Structure   Customer-driven Organization Structure

FIGURE 8.9
Tradition and customer-driven organization structures.

focus of the product. As Apple was able to capture more than 2.5% of the market in the first few months after the release, it has entered the growth phase very quickly acquiring loyal customers and a few innovators. However, in the growth phase, the sales of the product are almost stable with only slight gain in market share because of lacking value proposition and strong competition from giant market players that are already present in this industry such as Amazon and Google. The highlights of both the phases are provided in Table 8.2. It will be very interesting to see the next steps and strategies from Apple to accelerate its growth and outperform its competitors.

8.3.6   Attract and Retain Customers

Creating and maintaining long-term relationships with customers are critical regardless of whether an organization serves to B2C or B2B customers. Specific actions taken by a company can gain or lose the trust of customers. Organization chart and structure play a key role in managing relationships with the customers. There are two common variety of organization structures: traditional and customer-driven [24, Chapter 5] as shown in Figure 8.9.

In a traditional structure, executives are at the top followed by management, field staff, and customers at the bottom. In contrast, the

Comparison of HomePod Life-Cycle Stages	
Introduction	**Growth**
Characteristics - Low sales (less than 5 millions a year) - High advertising and customer acquisition costs - Very high competition from Amazon and Google - Loyal brand customers and niche innovators who are not satisfied with the music quality of existing products	*Characteristics* - Growing sales and slightly increasing profits - Market share is much lower than the competitors (less than 6% market share) - Still high advertising and customer acquisition cost because of not pioneering the market and higher competition from recent, smaller players such as Baidu - Early-adopters
Marketing Objectives - Creating product awareness by commercials and emphasizing the benefits to customers - Capture significant portion of market share from its competitors	*Marketing Objectives* - Increase market share - Become an industry leader in the home music industry
Strategies - Offer a product with enhanced features on better music quality and user convenience - Started with higher price ($349 USD) - Distribution and sale through Apple stores and website	*Strategies* - Only one product is offered - Cut the price from $349 USD to $299 USD to increase the market share - Distribution and sale is still through Apple stores and website because of slow market response

TABLE 8.2

Product stages of Apple HomePod, a product enabling smart residential buildings.

hierarchy and prioritization are inverted in a customer-driven organization. In this modern structure, the customers have the top most importance followed by the field staff, management, and executives. Field staff, management, and executives are interacting continuously with the customers to identify and satisfy the customer needs. Customers' preferences and choices are given higher priority during decision-making processes. As shown in the figure, customers have higher level of interaction and involvement with different parts of the organization at different levels. Although customer-driven structure motivates behavior to develop strong customer relationships, the structure may not necessarily be the best for every type of company [25]. Many major companies in the building controls industry exhibit a structure somewhere in between traditional and modern customer-oriented organization. In recent years, some companies have started to shift more toward customer-driven structure.

There are several stakeholders (installers, technicians, operators, engineering companies, and consultancy firms) in the chain before the final product is installed and sold to the end-customer. The customer, who is paying the bill, may not be aware of the product or the company who installs the product. For example, many people don't know the company and detailed specifications of the hot water system installed in their houses. Ways to attract and retain the existing customers and other stakeholders are:

- Excellent customer service

- Technical support including on-site visits

- Long-lasting support on the existing products

- Continuous interactions with customers after sale

- Responsive delivery of products and services

- Handling customer complaints and providing appropriate resolutions in a timely fashion

- Participation in trade show, conference, and exhibitions

- Road shows, learning academy, and collaboration with universities to attract both talent and future customers

8.4   Industry Characteristics

Companies are clubbed together based on the similarities in the features, products, services, customers, and usages. A set of companies in the smart building industry possesses certain characteristics that are beneficial to be

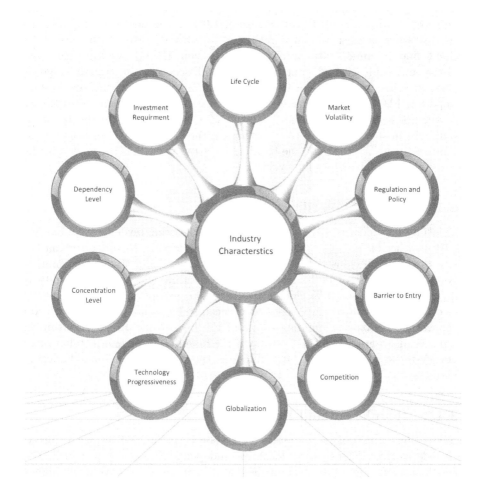

FIGURE 8.10
Industry characteristics.

understood by any business. Figure 8.10 lists the characteristics of building industry. The description on the characteristics is provided next.

8.4.1 Life Cycle Stage

Life cycle stage is divided into four major stages: start, growth, maturity, and decline. Growth rate, customer demand, and contributions of the industry to the economy and society determines the life cycle stage of industry. Life cycle stage is a variant of the type of the business. The HVAC manufacturing sub-sector, with expected growth rate of 1.9% in the next 5 years, is in the mature stage as most buildings in the US have HVAC systems and customers don't find an immediate need to replace their HVAC systems. However, BASs and sophisticated control systems sub-sector is in growth stage as these systems are still absent from many buildings [17]. The same observation applies to IAQ and other sensing capabilities. New techniques and technologies in the software/IoT world have started to change the landscape of smart buildings in short period of time. Although, the software and control systems industry is in growth stage, the building industry, overall, is considered in the early mature phase of its life cycle.

8.4.2   Market Volatility

Volatility is calculated using the change in market size, revenue, technologies, customer needs, and customer perception over 3–5 years. Revenue and market size are commonly used to quantify the changes and evaluate the market volatility. In most of building industry sub-sectors, the market volatility is quite low (less than 10% change over a year), except in the sub-sector of electronic or software components which need to be updated at faster interval than other systems in buildings. One of the biggest factors contributing to the revenue or growth of this industry is the manufacturing and installation of new systems, which is heavily dependent on the new construction and housing markets.

8.4.3   Regulation and Policy

Regulations and policy mandates correspond to the restrictions enforced by federal government, state, local and independent agencies to benefit the people, society, environment, and the country. Some states have much higher regulation and constrained policies than others. The agencies include AHRI (Air Conditioning, Heating, and Refrigeration Institute), ASHRAE (American Society of Heating, Refrigerating and Air-Conditioning Engineers), GAMA (Gas Appliance Manufacturers Association), IEEE (Institute of Electrical and Electronics Engineers), ICC (International Code Council), CEC (California Energy Commission), EPA (Environmental Protection Agency), and US DOE (Department of Energy), among others. Regulation level in building industry is medium and is expected to increase on several fronts. In this industry, there are many more regulations on the hardware components than software or control components, which is likely to change in the upcoming years as well. With new regulations that requires replacing the existing equipment or its component, the building industry is going to benefit from it.

8.4.4 Barrier-to-entry and Investment Requirement

Barrier-to-entry reflects the hurdles and challenges that the company overcomes to develop and introduce a new product to the market. In general, building industry has medium to high level of barriers to enter the market. Regulations, infrastructure development, initial capital investment, and domain expertise contribute to the barriers to enter the market. Manufacturing a large equipment requires substantial amount of initial investment and capability development because of the business nature and the liability associated with the equipment. For example, a gas-operated hot water system inside residences, if not designed, operated, and tested right, can cause damage to property, assets, and people lives. Depending on the channel strategy, a distribution network may also be needed. In contracting and services business, low investment is needed but practical domain expertise is needed with continuous learning. Therefore, barriers to entry in the servicing, contracting, and control business are at the medium level. To give an idea, there are 107,650 businesses [23] in the HVAC contracting sub-sector while 1,462 businesses [33] are currently present in the HVAC manufacturing sub-sector.

8.4.5 Competition and Concentration Level

Competition ranges from medium to high in the buildings industry. Large equipment and control manufacturers purchase large number of components while sharing common manufacturing facilities and suppliers. These manufacturers have invested significantly in setting up a capability and infrastructure that are required to design and develop the products. Over the years, the manufacturers have also acquired and cultivated the skill sets of people for their products. Entering in this market has medium barrier level requiring large initial investment. Therefore, the competition level is neither low nor extremely high in these sub-sectors. Conversely, in the contracting, servicing, and control businesses, the competition is high because of low initial investment and quick returns.

In most markets, there are many medium/large players instead of one single dominant player. Although many players serve in the vertical market, their market share vary quite a bit in different building sub-sectors of the market. As an example, UTC/Carrier, Goodman/Amana, Trane/American Standard, Lennox, Rheem, York and Nordyne dominate the residential HVAC equipment market, but these companies are not the market leaders in residential HVAC controls [15]. Because every company capture small portion of the total market share and there are many small businesses in the contracting business, the total market concentration level is considered low [23].

8.4.6 Globalization

Import/export of products/services and presence of companies in different countries decide the level of globalization. Building industry is fragmented from globalization perspective. Most companies in services and contracting businesses rely on local contractors while large manufacturing have global presence with manufacturing facilities and operational presence in different parts of the world. There is low import/export from/to other countries. Therefore, overall globalization is low to medium in this industry, tending to lean toward medium in upcoming years.

8.4.7 Technology Progressiveness

Historically, building industry is slow in adapting and incorporating technological changes. In recent years, the technologies in the systems have changed significantly in every part of design, manufacturing, service, contracting, and customer-service businesses. Several products exist in the market leveraging the IoT, Internet, cloud, automation, smart devices, and digital technologies in many forms and shapes. Examples of products in residential market are iManifold [9], Comfort Guard [11], AirAdvice [2], FooBot [6], and smart connecting thermostats such as Nest [7], Ecobee [5], and GLAS [16].

8.4.8 Dependency Level

Dependency level indicates how the revenue and market share of the industry is dependent on external factors and the success of other industries. Lower is the dependency level, the lower are chances of the industry to be impacted by the surges in other companies. Trends in construction industry heavily shape the trends in smart building industry. Sensors and several components in the embedded controls in the building industry are shared with automotive and electronics industries. Therefore, the dependency level is medium in this industry, which is expected to remain same and increase in the future. Table 8.3 summarizes the characteristics and their position in the building industry.

8.4.9 Industry Structure using Porter's Five Forces

Porter's five forces is another way to examine and evaluate the industry structure and competition in a business category. The five forces are threat of new entrants, threat of substitutes, bargaining power of customers, bargaining power of suppliers, and competitive rivalry. Another force introduced or proposed to this model is power of complement providers. These forces and the risks associated with the forces are described next.

Buildings Industry Characteristics	
Characteristics	**Status/Trend**
Life Cycle	Mature
Market Volatility	Low
Regulation and Policy	Medium
Barrier to Entry	Medium-High
Competition Level	Medium-High
Globalization	Low
Technology Progressiveness	Medium
Concentration Level	Low-Medium
Dependency Level	Medium
Investment Requirement	Medium-High

TABLE 8.3
Characteristics and attributes associated with the building industry.

1. Bargaining power of suppliers (Moderate): This power reflects the negotiating power and pressure that the suppliers possess when working with the company. Higher is the bargaining power, the higher are the risks to the company. This power is at moderate level because there are only limited suppliers available in the market. Moreover, the components or parts in the building system are also being used by other industries, e.g., sensors and microprocessors in automobile sector. Other sectors use those components at much larger scale than the building industry. Thus, when the demand is high in other sectors, the supplier's priority may shift accordingly.

2. Threats of substitutes (Low): This power corresponds to the options that can act as substitutes and replace the existing products. Higher the number of substitutes, the greater is the risk to the company or industry. There are not many products that can act as a substitute for building systems. Therefore, the threats are very low in this

category. For instance, the only possible alternatives of HVAC systems are fans, natural cooling/heating, or living in climate that does not require much air conditioning or heating, e.g., Seattle.

3. Threats of entry barriers (Low): Developing a full ecosystem of building systems requires substantial capital investment as discussed in Section 8.4.4. This requires specific expertise/skill sets, knowledge and capabilities such as labs, testing equipment. Therefore, entry barriers in this industry are high.

4. Bargaining power of buyers (Moderate): For most building products, customers initially have many options to buy a product from several companies. However, each product has a steep learning curve. It means that the stakeholders (installers, operators, homeowners, etc.), who have bought a product from one company are likely to stick with them unless there are strong reasons to shift. Basically, switching costs associated with the products are not just the knowledge transfer/learning but also the capital investment in changing the products. For all these reasons, the bargaining power of buyers are moderate.

5. Industry rivalry (High): There are many large companies in this industry because of the total market size. However, no single company dominates the market in every sub-sector of the industry. One company may be large in one portion of the market or one specific product category, while the other company may be larger in the other category.

6. Power of complement providers (Low): This power corresponds to the complement goods (or the companies with complement goods) that add value to the existing products, e.g., software applications are compliment goods for a computer. There are numerous small companies that provide additional services, e.g., software solutions to improve building performance or detect problems with the equipment. Typically, large companies aim to incorporate such services in their upcoming products. Doing so reduces the power of complement goods. Moreover, third-parties (or complement firms) have to invest continuously in maintenance and new ideas to stay interoperable and up to date with the main products from large companies. For instance, if Microsoft releases a new version of Windows, third-party software companies have to release a new software application that can run on the new version of Windows. Therefore, the power of complement providers is low. Propriety and closed nature of businesses also creates huge dependency for the complementary providers.

Basically, the industry reaps the benefits of high entry barriers, low threats of substitutes and complementary providers, and moderate bargaining power of suppliers and buyers. Although, the industry has slower growth rate with high rivalry, there are many potential opportunities for the companies if they can adapt to the changing environment. Therefore, from the competitive forces model, the industry continues to lean more toward a profitable industry.

8.5   Marketing Challenges and Opportunities in Building Controls

Building controls offer technical and management challenges that need to be considered and overcome for both initial and long-term success of products. Control products—after tackling the challenges in right fashion—can also provide competitive edge for an entire business unit. This section discusses the challenges and their corresponding opportunities:

- Uniqueness: Every building is unique with distinct requirements, layout, thermal characteristics, orientation, weather constraints, and features. Buildings' uniqueness creates a challenging situation to develop a quality product quickly that targets all the variations. It means that a control product should consider the maximum possible variations during the development, testing, and validation phases.

- Controls Applicability: There are a number of HVAC system types with individual modifications and configurations as shown in Chapter 1. For example, if an occupancy sensor is not available, the control algorithms may use a schedule which is not a real-time indicator of occupancy. Similarly, there can be almost half a dozen types of duct static pressure reset strategies for an AHU-VAV system. The master control algorithm designed for a system should be robust enough to handle different variations and configurations.

- Application Deployment Time: Time required to deploy or upgrade the existing control applications should be very low to reduce the installation cost.

- Installation Cost: Similar to the application deployment time, the total installed cost must be reduced. This could be accomplished by designing a hardware and software that require minimal manual interventions.

- Stakeholders: As there are many stakeholders in the buildings industry discussed in Chapter 7, control engineers and product managers should prioritize the stakeholders and address the concerns of several stakeholders. A crude weighting criteria can be formulated to maximize the output.

- Turn-around and Upgrade Pathway: Any control solution must be ready to face unexpected bugs and issues when implemented in buildings. It is important to ensure that there is a quick pathway to release and deploy updates to fix the problems. The same or similar pathway can be adopted for new control solutions that outdate the existing controls.

• Benefit: In short, a new, effective control solution requires low investment while delivering the results with high impact in a short period of time. Quicker development is possible because of fast development cycle and less dependency on external factors.

8.6   Growing Trends and Environmental Factors

Capturing and tracking the recent trends and environmental factors are essential for long-term success of a company. Good companies realize the changing environment and capitalize on the opportunities to gain a competitive advantage. A few macroeconomic forces and the corresponding long-term trends have been influencing the building industry over the past decade.

First, primary stakeholders (building owners and occupants) are getting more educated on the importance of systems (lighting, air quality, air conditioning) inside buildings; they are highly interested in learning about the systems and their environmental impacts such as greenhouse emissions and efficiency. At the same time, new energy standards and building codes (considered as regulations) are being developed, which are strongly requested by the customers. This imposes a risk to the company as the products need to be enhanced and new capabilities need to be developed with additional investment to satisfy the customer requests. In contrast, at the same time, this situation can be thought of as an opportunity because the company can gain higher market share if such products can be developed in a timely fashion to outpace the competitors and increase market entry barriers for new players.

Second, the customers and other participants in the value chain— technicians, installers, operators, facility managers, etc.—are becoming comfortable with smart devices and new technology such as smart phones, tablets, wireless sensors. This is a huge opportunity as the company can shift their strategy from hardware-focused solutions to software-focused solutions, which would not have been effective a few years ago. Doing so has several major benefits: competitive edge, higher profit margins on controls and software applications, and potential subscription-based revenue streams. With strong marketing management and product development while monitoring these external trends, it is necessary to develop a strategy and position the company in the market continuously. It is possible that several major decisions may need to be taken for re-positioning and updating the firm-level strategy.

Third, the Internet, social media, emails, and cloud have changed the way everyone performs their day-to-day operations. With a smart device in hand most of the times, information can be communicated to the stakeholders effectively and efficiently. Prevalence of these tools also enables the users to provide feedback to the companies in real-time. Use of such tools and technologies opens new doors of opportunities for controls and software companies.

Recently, COVID-19 has not only impacted many industries (some more than others) but also revealed the gaps and shortcomings of several sectors. COVID-19 had changed the (1) way people work/live/interact in the buildings, (2) functionality, utilization, and purpose of several building types, (3) the equipment and devices installed in buildings, (4) market demand of certain building types, (5) processes and the activities that have been carried out before and after the construction, (6) regulation standards and mandates imposed by the government in public/private places, and (7) priorities and preferences of different stakeholders (primarily the occupants and building managers) associated with buildings. As an example, many companies have started to offer work-from-home as a permanent option for both existing employees and new hires. Air purification systems and advanced filtration techniques have been installed inside buildings to clean the air frequently and improve the IAQ. Ventilation to bring a higher level of outdoor air have been an important and essential part of the HVAC system inside the buildings. Many changes along these lines took place to resume the activities in a "new normal way." Some companies have released new products while others have changed their business models and processes. For example, AI-based mask detection products have been launched into the market to ensure safe opening of work places. Other examples are portable air purifiers, HVAC embedded air filtration systems, remote system monitoring, and asset tracking with temperature monitoring. This raises several important questions: how long this change will last or how much change will be permanent in day-to-day operations or how much change will fade away with time? Based on the responses to such questions and changing market trends, new market opportunities are expected to emerge. As a result, the market players (new and existing) who are able to adapt and capitalize on these opportunities are likely to arise, survive, and thrive.

"Measure what can be measured, and make measurable what cannot be measured."

Galilei, *Galileo*

1. Understanding the broader definition of a product (or a set of products) from marketing perspective is necessary for communicating and delivering a strong value proposition.

2. Marketing management includes developing strategies on product, marketing, targeting, pricing, positioning, growth, and customer relationships.

3. Common variables to segment the buildings market are stakeholders, building type, climate, geography, building size, customer usage, user proficiency, loyalty status, family size, income level, generation, education, building condition/age, and user status.

4. SWOT analysis provides an extremely useful framework to evaluate internal (strength and weaknesses) and external (threats and opportunities) attributes associated with a product, brand, or company.

5. SWOT analysis and competitive analysis must be performed at regular interval for better positioning and competitive advantage of the products.

6. Growth can be accomplished by increasing the customer base, new products or adding value to the existing products, widening the portfolio in different markets, and improving the customer perception of the brand.

7. A good screening process eliminates a bad product idea in earlier stages to minimize the wastage of resources and time.

8. Evaluating the timeline and current product stage (introduction, growth, maturity, and decline) guides the next steps in developing product and marketing strategies.

9. Industry trends and characteristics are needed to recognize the challenges and opportunities in an industry for a company and its competitors, and thus the true potential of a product by selecting an appropriate set of strategies.

10. Industry characteristics are life cycle, market volatility, regulation and policy, barrier to entry, competition level, globalization, technology progressiveness, concentration level, dependency level, and investment requirements.

11. Increase in customer education, increase in technology adoption, use of new technologies (Internet and digital tools) are growing trends—that can be capitalized on—in this industry.

Bibliography

[1] AHR EXPO. https://ahrexpo.com/. Accessed: 2020-03-22.

[2] Air Advice. https://www.airadviceforhomes.com/. Accessed: 2020-03-22.

[3] Apple Worldwide Developers Conference. https://developer.apple.com/wwdc20/. Accessed: 2020-03-22.

[4] ASHRAE Conference. https://www.ashrae.org/conferences. Accessed: 2020-03-22.

[5] Ecobee. https://ecobee.com/. Accessed: 2020-03-22.

[6] Foobot. https://foobot.io/. Accessed: 2020-03-22.

[7] Google Nest. https://nest.com/. Accessed: 2020-03-22.

[8] HomePod. https://www.apple.com/homepod/. Accessed: 2020-03-22.

[9] iManifold. https://imanifold.com/. Accessed: 2020-03-22.

[10] RESNET Conference. https://www.resnet.us/conference/. Accessed: 2020-03-22.

[11] Sensi Predict (previously known as Comfort Guard). https://sensi.emerson.com/en-us/products/sensi-predict. Accessed: 2020-03-22.

[12] Building Connect+ Mitsubishi Electric Trane HVAC US. https://www.esmagazine.com/articles/100186-building-connect-mitsubishi-electric-trane-hvac-us, Feb 2020. Accessed: 2020-03-22.

[13] Anna Amir. Security Alarm Services in the US–IBISWorld Industry Report 56162. Technical report, IBISWorld, Aug 2019.

[14] L Callaway and N Strother. Smart Thermostats: Communicating Thermostats, Smart Thermostats, and Associated Software and Services: Global Market Analysis and Forecasts, 2015.

[15] M.C. Baechler C.E. Mertzger, S. Goyal. Review of residential comfort control products and opportunities. Technical report, Pacific Northwest National Laboratory, Dec 2017.

[16] Johnson Controls. GLAS. https://glas.johnsoncontrols.com/. Accessed: 2020-03-22.

[17] Dan Cook. Electronic Design Automation Software Developers in the US–IBISWorld Industry Report OD4540. Technical report, IBISWorld, Mar 2019.

[18] Toshiba Corporation. Toshiba and United Technologies Sign New Agreement to Grow Joint Venture. https://news.toshiba.com/press-release/corporate/toshiba-and-united-technologies-sign-new-agreement-grow-joint-venture. Accessed: 2020-03-22.

[19] Jack Curran. Elevator Manufacturing in the US–IBISWorld Industry Report OD4684. Technical report, IBISWorld, Aug 2019.

[20] Heidi Diehl. Building Lighting Control System Manufacturing in the US–IBISWorld Industry Report OD4518. Technical report, IBISWorld, Dec 2018. Accessed: 2020-03-22.

[21] Heidi Diehl. Fire & Smoke Alarm Manufacturing in the US–IBISWorld Industry Report OD4475. Technical report, IBISWorld, Dec 2018. Accessed: 2020-03-22.

[22] Johnson Controls Inc. Johnson Controls and Tyco complete merger. https://www.johnsoncontrols.com/media-center/news/press-releases/2016/09/06/johnson-controls-and-tyco-complete-merger. Accessed: 2020-03-22.

[23] Kevin Kennedy. Heating & Air-Conditioning Contractors in the US–IBISWorld Industry Report 23822a. Technical report, IBISWorld, Dec 2019.

[24] P. Kotler and K.L. Keller. *Marketing Management.* Pearson Education, Limited, 2015.

[25] Ju-Yeon Lee, Shrihari Sridhar, and Robert Palmatier. Customer-centric org charts aren't right for every company. *Harvard Business Review,* 06 2015.

[26] John C Mankins. Technology readiness levels. *White Paper, April,* 6:1995, 1995.

[27] Jeremy Moses. Electronic Access Control System Manufacturing in the US–IBISWorld Industry Report OD4477. Technical report, IBISWorld, Dec 2018. Accessed: 2020-03-22.

[28] US Department of Commerce and NOAA. NWS JetStream – Climate, Aug 2019.

[29] U.S. Bureau of Labor Statistics. Table 1101 Quintiles of income before taxes, Aug 2016.

[30] Ryan Roth. HVAC Service Franchises in the US–IBISWorld Industry Report OD5583. Technical report, IBISWorld, Aug 2019.

[31] Ryan Roth. Automatic Fire Sprinkler System Manufacturing in the US–IBISWorld Industry Report OD5509. Technical report, IBISWorld, Jan 2020.

[32] U.S. Energy Information Administration (EIA). *Annual Energy Outlook 2019*. DIANE Publishing, Jan 2019. Accessed: 2019-06-30.

[33] Gordon Zheng. Heating & Air Conditioning Equipment Manufacturing in the US–IBISWorld Industry Report 33341. Technical report, IBISWorld, Dec 2019.

[34] Gordon Zheng. Heating & Air Conditioning Wholesaling in the US–IBISWorld Industry Report 42373. Technical report, IBISWorld, July 2019.

9

Financial, Cost, and Supply Chain Management

> "A penny saved is a penny earned."
>
> — Benjamin Franklin

Managing the cost and finances are equally vital as other functions (marketing, product development, etc.) of a successful business. Setting, achieving, and improving on financial targets are usually the top most goals of businesses as many businesses strive for long-term sustaining growth while making smart decisions. This methodology applies not only to a company but to an individual project, a product, or a group of products and projects. Asking the right financial questions during project evaluation, product development, and pricing strategies can result in better project/product selection. Product and project are used interchangeably in this chapter. The main focus of this chapter is to answer two key questions:

1. How financial evaluations can influence or help in the selection of a product from monetary perspective only?

2. How can the overall cost of a product be reduced by assessing alternate solutions and implementing different measures?

To answer these questions, a few financial concepts need to be understood, which are explained next in this chapter. Supply chain management is also explained in this chapter because it is a powerful way to reduce the cost and improve the company operations. It is important to note that the topics covered in this chapter are related to the cost and finances of a project, not an entire company.

9.1 Financial Concepts

9.1.1 Rate of Return and Interest Rate

Rate of return and interest rate are very similar to each other as both represent the rate at which an investment or an amount of money gains value over time.

The main difference between them is the type of investment they are used for. Interest rate corresponds to the extra amount paid on a loan every year or month, while rate of return corresponds to the gain received on an investment. For example, suppose that you have invested $100 in an internal project or the stock market. If you have sold the stock to $110 after a year, your gain is $10 yielding a 10% annual rate of return. Rate of return is mostly used in the context of this chapter.

9.1.2 Present Value vs. Future Value

Present value corresponds to the value of an asset or cash equivalent on today's date while future value represents the equivalent value in some future date. These terms are also heavily linked to the time value of money concept [14, Chapter 28], which states that the current money is worth more than the same amount of money in the future as one can accrue interest on the money. In other words, the value and the purchasing power decrease over time if you have the same amount of money. Present (PV) value and future value (FV) can be expressed as:

$$FV = PV(1+r)^n, \qquad (9.1)$$
$$PV = \frac{FV}{(1+r)^n}.$$

Rate of interest or standard rate of return over a time period n is represented by r. Future value indicates the value of cash invested today in the future. The concept and the terms are very important for understanding finances as they provide a relationship between the present value and future value at different times. Since the concept is abstract, the following example should help understand the concept.

Example: If someone asks you $100 today and promises to return you $120 next year, would it be worthy? Assuming the inflation rate and rate of return is 10%, the answer is yes because the future value (i.e., $110 calculated from Eq. 9.1) is lower than the amount offered by the person.

However, if the same person promises you to return $150 in 10 years instead, would the deal be worthy for you? The answer is no because the future value after 10 years (Eq. 9.1) is much higher than what was offered by the person in 10 years. Figure 9.1 illustrates the future value of this example as a function of time using Eq. 9.1.

9.1.3 Inflation

Inflation is an act of increasing the prices of services and goods. Inflation rate is the rate at which inflation happens. Inflation results in reduced purchasing

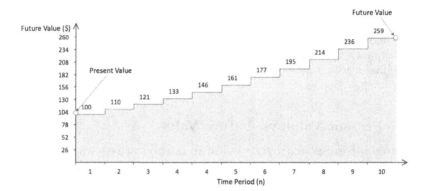

FIGURE 9.1
Future value as a function of time and current value.

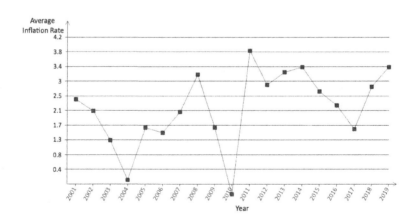

FIGURE 9.2
Annual average inflation rate in the US over the last two decades.

power as the value of money decreases. Figure 9.2 shows the historical inflation rate in the US over the past several years [2].

If the growth of a company is less than the inflation rate, it means that the company is losing its worth in present value. The same principles apply to individual products. Purchasing power affects both consumers and businesses. If an investment or nominal return on a product in one year is 10% with inflation rate 2.5% in the year, the real gain or investment return is 7.5%. Therefore, incorporating inflation in the financial equations is beneficial. The sale and overall profit margin for a product should be higher than the inflation rate to sustain the business provided all other factors are the same. In fact, a prosperous business aims to outpace the growth of industry in addition to the inflation.

9.2 Project Selection–A Financial View

This section focuses on making the product or project related decisions while comparing their financial situation. Suppose that there are multiple products or projects that a business leader has to choose from, e.g., creating a new hardware controller, launching the existing controller to other markets. The factors and performance metrics that need to be evaluated for making the best decisions for the company are described in this section.

9.2.1 Return on Investment

Return on investment is the simplest form to evaluate the outcome of an investment. It determines the gain or losses for the investments. Return on investment (ROI) is defined as the percentage ratio of gain or losses and the amount invested:

$$\text{ROI} = \frac{\text{Gain (or loss)}}{\text{Invested Amount}} \times 100\%. \tag{9.2}$$

Higher the ROI, better is the project or product. ROI is simple and quick to calculate. One of the biggest drawbacks of ROI is that it does not consider the time value of money. Therefore, it is not precise and accurate for long-term projects. Despite its disadvantages, it is a commonly used metric because it is simple and provides a crude/fast analysis of the situation.

9.2.2 Simple Payback Period

Simple payback period (SPP) corresponds to the time it takes to get back the money that was invested is a project or spent in purchasing a product. SPP is also an easy and quick-to-calculate financial metric associated with a project.

SPP is defined as the ratio of invested amount and the average return per year:

$$\text{SPP} = \frac{\text{Total Invested Amount}}{\text{Average Yearly Return}}. \tag{9.3}$$

Similar to the ROI, SPP does not consider the time value of money. Furthermore, the fluctuations in the return during different years, especially after the payback period, is not taken into account.

Example on ROI and SPP: Suppose that a building owner has decided to upgrade its existing control system to an advanced DDC (Direct Digital Control) system that has additional sensors and controllers with energy-efficient control algorithms. The total initial investment of this project is $60,000 USD. The building energy bill $100,000 USD. The new system claims to save the energy by 20%, resulting in total energy saving of worth $100,000 USD over 5-year period assuming a flat electricity price. In this time period, the ROI turns out to be 67% using Eq. (9.2), which is a good number. Similarly using Eq. 9.3 for the $20,000 USD savings every year, the SPP turns out to be 3 years. Note that several other ongoing costs such as maintenance, operation, and training costs are not taken into the calculations. Therefore, ROI and SPP are rough but important metrics to analyze the investment.

In the above example, it is not clear how a 5-year period is chosen for analysis purposes. Although a shorter period is preferred by businesses, it is not possible to generate acceptable profits in short time periods. This depends on the product type, product life cycle, and its developmental time. If the time period is chosen too large, the product may not generate any profit in the first few years. Thus, choosing a time period is a trade-off between several short-term and long-term factors.

9.2.3   Net Present Value

NPV (Net Present Value) combines the present value of all the investments, losses, and gains when they occur. NPV uses the time value of money concept to determine the values. NPV is expressed as:

$$NPV = -\text{Initial Investment} + \sum_{i=1}^{i=n} \frac{\text{Gain(Losses)}_i}{(1+r)^i}, \tag{9.4}$$

where r is the standard rate of return, i corresponds to the ith time instance of total n time periods, and $\sum_{i=1}^{n} \frac{\text{Gain (Losses)}_i}{(1+r)^i}$ represents the summation of the present value of all the investments, gains, and losses over the entire time period ending at nth instance. Investments and losses are subtracted while the gains are added to the equation. The concept can be used to decide the best decisions (amount projects or within a project) from financial point of view. The following example illustrates the concept.

Year	Supplier A	Supplier B	Supplier C
0	($50,000)	($70,000)	$0
1	($15,000)	($25,000)	($30,000)
2	($15,000)	($25,000)	($30,000)
3	($15,000)	$0	($30,000)
NPV	($88,656)	($114,582)	($77,313)

TABLE 9.1
NPV (in USD) on contracting cost from Supplier A, B, and C for 3-year long product.

Example on NPV: Let say that you are developing a control software that requires purchasing or licensing libraries or a software from an external company. You are looking for a purchase/maintenance contract for 3 years. There are three available suppliers: Supplier A, Supplier B, and Supplier C. Supplier A charges $50,000 USD upfront and $15,000 per year afterward for the next 5 years. Supplier B charges $70,000 upfront and then $25,000 per year for only two-year life of the contract. Supplier C does not charge anything upfront. Instead, it charges $30,000 USD every year regardless of the number of years. Table 9.1 shows the investment cost involved with the suppliers for 3-year time period. Numbers in braces () indicate the cash flowing out of the company.

Assuming the annual rate of return is 8%, NPV for all three suppliers are provided in the last row of the table. Lower is the value in the round brackets, the less is the amount of investment needed for your company to develop the product. It is clear from the table that Supplier C has the lowest cost because of no upfront charges despite its higher yearly maintenance cost. Supplier with next lowest cost is Supplier A followed by Supplier B again because of lower initial cost with Supplier A as compared to Supplier B. The initial cost is playing a huge role in this scenario because the total time duration is only 3 years, which is very small. The next question is: what happens when the expected contract time and product life-cycle is much more than 3 years? With the same cost structure as mentioned above, the NPV of all the suppliers as function of time is shown in Figure 9.3.

As the number of years increases shown in Figure 9.3, the NPV of the supplier cost changes. Depending on the length of the contract and product life-cycle, the selection of best supplier changes. Over a long time period, Supplier B is the obvious choice while Supplier C is the obvious choice for a

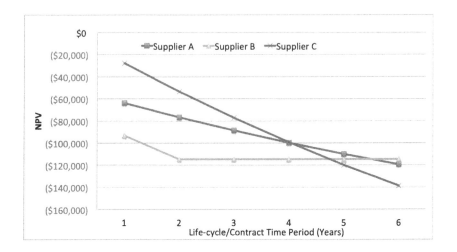

FIGURE 9.3
NPV on contracting cost from three suppliers as a function of product life-cycle/contract.

short-term project. Supplier A is a good choice when the product duration is in mid-range (4–5 years). Supplier A is also a good choice when the project duration is between 3 and 10 years but there is too much uncertainty during that time period because of less variance and risks.

NPV concept is also an option to (1) calculate the break-even time period or a payback time period that accounts for the time value of money concept, and (2) prioritize the list of projects or products within the company. For example, if your department or company has a budget to invest $20 million dollars with the several product options: A, B, C, D, E, F, and G. The initial investment and NPV of the projects are shown in Table 9.2.

Sequential steps involved in prioritizing and selecting the products for development are:

1. Filter out the projects that do not meet the criteria for business reasons, e.g., misalignment with corporate strategy, non-feasibility, and failure of similar products in the past.

2. Determine the initial capital cost for each selected project along with its sale projections and the total expenditure budget.

3. Calculate the NPV for every project using Eq. (9.4).

4. Calculate the financial priority ($\frac{\text{Initial Investment}}{\text{NPV}}$) of the projects.

5. Rearrange the projects in the descending order of their priority.

Product	NPV (in $ million)	Initial Investment (in $ million)
A	$1	$2
B	$1.5	$6
C	$10	$12
D	$5	$10
E	$2	$5
F	$2	$3
G	$3	$4

TABLE 9.2
NPV and development cost of multiple products.

6. Calculate the combined total initial cost starting with the project that has the highest priority.

7. In the prioritized list, mark the cut-off line where the total initial cost is closest but less than the budget, i.e., Total Cost $\leq$ Expenditure Budget.

8. Select the projects above the cut-off line.

Following the above process for the example mentioned in Table 9.2, a list of selected projects and their corresponding calculations are shown in Table 9.3.

As discussed in this section, NPV is a relatively precise and accurate method to determine if an amount invested today is worth it. It incorporates multi-year transactions and cash flows with varying rate of returns because the future cash flow can be risky. However, there are still some areas where the NPV concept lags. One of the main disadvantages of NPV is to determine the time period (number of years) and the standard rate of return for analysis ignoring future projections. Other disadvantages of NPV are improper handling of the projects of unequal length/sizes and obtaining the time-line of cash flows. There are several other metrics that evaluate different variables associated with a product or project. Other methods or metrics are IRR (internal rate of return) [7], MIRR (modified rate of return) [13], and benefit-cost ratio [9].

Project	NPV (in $ Millions)	Initial Cost (in $ Millions)	Priority	Total Cost (in $ Millions)	
C	$10.0	$12	0.83	$12	⎫
G	$3.0	$4	0.75	$16	⎬ Selected Projects
F	$2.0	$3	0.67	$19	⎭
A	$1.0	$2	0.50	$21	
D	$5.0	$10	0.50	$31	
E	$2.0	$5	0.40	$36	
B	$1.5	$6	0.25	$42	

TABLE 9.3
Prioritized cost and optimal selection of products using NPV.

9.3 Cost Management

After analyzing the market, competition, customer needs, and financial constraints, the product for development is crudely decided and finalized. Based on the competition, customer demand, and other relevant factors discussed in Chapter 8, the product pricing strategy is mostly settled. The next major step in the entire plan is to manage and reduce the overall cost of the product. Figure 9.4 shows a fundamental relationship between profit, price, and cost. The lower the total cost, the higher is the total profit for a given price of products. Usually, there are constraints on prices, which are heavily driven by external factors. For instance, it is very hard to sell a residential thermostat at $500 if the thermostat provides similar value and features as other existing thermostats, which are currently sold for $150–$300. The cost of a product is also dependent on the internal factors. In fact, cost is a measure of internal inefficiencies including its supply chain management. This section discusses the techniques to reduce the total cost.

9.3.1 Cost of Goods

An obvious method to reduce the cost of a product is to decrease the cost of goods sold (COGS). The COGS sold corresponds to the amount of money paid to vendors for supplying raw components. Therefore, the cost of goods applies only to tangible products and physical components, e.g., sensor, embedded controller, external control hardware, server, and CDs/USB-drive for distributing control software and algorithms. It is obvious that the cost of

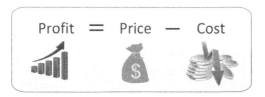

FIGURE 9.4
Basic financial equation connecting price, cost, and profit.

goods is much higher for hardware-based control solution than the COGS of only control algorithms or software applications. Therefore, reducing the cost of goods of physical controllers is important to yield high profit margins.

9.3.1.1 Financial Factors

The following factors are needed to be considered while selecting a supplier or the goods for a hardware product:

- Initiation Cost: It refers to the initial investment before any product is procured or purchased, e.g., costs to setup contracts or perform preliminary development work.

- Purchasing Cost: This cost reflects the amount paid to purchase the goods, e.g., the price of micro-processing chip for the controllers.

- Maintenance Cost: It is related to the reoccurring cost to maintain the components in the product. Suppose a company had procured the product micro-controllers for 3 years in advance because of hike in demand and market price forecasts. If the company still incurs any cost that need to be paid to the vendor, it will be considered under maintenance costs. An example is the licensing cost of a software that is required to program the micro-controllers or modify the existing software on the micro-controllers.

- Retirement Cost: It corresponds to the cost associated with retiring components from the product. Usually, this cost is taken into account when there is high volatility in technologies. Software solutions are common examples in this area. Hardware control products have much longer life-cycle than software applications, i.e., the hardware components don't change as frequently as software components. Therefore, the retirement cost is weighted less in the overall picture.

- Replacement Cost: Replacement cost is examined when the hardware components are expected to be replaced by other components. The replacement components can be from the same vendor or the vendor's competitor. This cost has a visible affect when many components are expected to contribute highly to the overall product cost. Trends on the prices of components are highly valuable in the analysis, e.g., decreasing cost of Wi-Fi chips (less than a couple of USD [8]) and CAT5 cables (plenum rated cable priced at approximately 10 cents/ft [10]) may replace the existing RS-485 port with an Ethernet port.

- Credit Policy: Credit policy contains the terms and conditions that is acceptable to the vendor. Credit policy includes the credit limit, payment time period, and penalties associated with missing payments. Higher is the credit limit and payment time period, the less is the cost of the components because of the time value of money concept and higher cash-flow.

- Discount: Discount on the components is another way to lower the cost as some vendors may offer discounts for bulk purchasing, advance payments, and no-credit payments.

- Government Incentives: State and federal agencies may offer packages and rebates to encourage the use of environment-friendly components and products. The incentives could be given to the customers or the companies. This category is not highly applicable at the component level, but rather at the system level, e.g., solar panels and electric vehicles.

After understanding the aforementioned costs and estimating the requirement of components, the time value of money concept with NPV (in Section 9.2.3) can be used to compare multiple options. To select multiple components in a product, one must be careful about the number of suppliers. There may be additional costs while coordinating with high number of suppliers and tracking the interoperability between the components supplied by different vendors. The coordination cost is based on the quality of vendor management/tracking system inside the company. For instance, Apple has more than 170 suppliers for different components in iPhone [11]. Despite the large number of suppliers, Apple is able to manage and deliver products through its effective supply chain management. For developing a new product when a company does not have a sophisticated vendor management system, it is a good practice to limit the number of suppliers.

9.3.1.2   Non-financial Factors

Besides the financial costs mentioned above, there are several non-financial factors that can indirectly alter the product cost although they are not easy to quantify. The non-financial factors are most likely to affect the operational expenses instead of the cost of raw materials. The factors include both tangible and intangible attributes associated with the supplier such as existing/past relationships, supplier flexibility to accommodate exceptional cases, supplier portfolio, return and inventory policies, and supplier payment system. As an example, good relationships with its suppliers lead to faster results and quick resolution in case of conflicts. This reduces the overall cost including the operational costs.

9.3.2 Operational Expenses

Operational expenses relate to the maintenance, business, development, and other non-tangible expenses—the expenses associated with running the businesses that do not include the cost of goods or other hardware expenses. Only operational expenses associated with the development of building control products are discussed in this section. Operational expenses are relatively high for developing a software-based controls product as opposed to a hardware product. Below are a few relevant costs (not exhaustive) contributing to the total operational expenses of control software products:

1. Research and development costs

2. Marketing and advertising costs

3. Installation cost

4. Maintenance and post-sales costs, e.g., training, support, and customer service

5. Administrative cost

6. Other costs such as entertainment, sales, travel, legal, and supplies.

In general, there are many common ways to reduce or cut the operational expenses. These methods vary from reducing the printing to laying off the workers; these methods are described in the articles [4, 3, 19]. From product development and delivery point-of-view, a few major expenses and possible techniques to reduce them are being touched upon next:

9.3.2.1 Automation

If a task(s) in the company is repetitive and time-consuming, automation is a quick way to finish the task in less time. Automation also reduces errors and helps employees improve their productivity and morale. A simple example of automation in most businesses is found in setting up a meeting room. An automatic confirmation/rejection is sent to the organizer immediately after the meeting room is requested. This example illustrates time/hassle savings for not only the participants and organizer but also the company as the company may not need to hire an individual (or hire less individuals) in managing the meeting rooms. Automation can affect many operational expenses in a positive way.

A good example of automation in building controls is automated testing of its hardware and software products. Chapter 5 discusses the ways of testing the system in an automated fashion. Automated testing benefits the (1) development team by reducing the time and errors; (2) company by delivering high quality products, reducing the operational expenses, and developing the products faster; (3) the customer-service team in reducing the customer-support expenses because of reduced complaints and calls; and (4) customers by reducing the number of complaints, increasing the perceived value, and delivering a higher sense of satisfaction. Automation is applied to a variety of internal processes, e.g., finance operations can be automated to deliver the payments on or before the due date to avoid penalties and increase cash flows.

9.3.2.2 Outsourcing

Outsourcing corresponds to providing a part of work to people outside the company (locally, domestically, or overseas). Reasons and motivations for outsourcing may include (1) reduction in operational expenses, (2) faster delivery time by pulling in additional resources, (3) unavailability of skill-sets in the company, (4) temporary or intermediate assignment that may end in short-time, and (5) in-house product development does not align with the corporate strategy. Outsourcing the work overseas—in which the people with similar skill sets but less salary because of the lower cost-of-living—is commonly done to reduce the operational expenses.

As the development cost is very high in developing software or control algorithm products, outsourcing the work is an effective way, if done right, to reduce the total product cost. Besides the advantages of outsourcing, there are several disadvantages including limited control and oversight, work reliability, transfer of knowledge and learning experience, protecting intellectual property, etc. Outsourcing is not always effective in every case, e.g., cross-country collaboration may not be beneficial if significant hardware and laboratory setup is needed to perform the development work. Suppose a company develops a remote BAS service to monitor the health of building systems. The software or a remote person notifies the building owner in case of problems or potential future issues in the system. If the control software is implemented on critical facilities, outsourcing may not be a viable option because of legal, privacy, and security concerns although most of the work can be independently performed at remote locations overseas. These factors should be taken into account while deciding not only whether outsourcing is beneficial for the company but also the right level of outsourcing.

9.3.2.3 Workflow and Processes Optimization

Workflow and processes are part of every business and product life-cycle management, e.g., planning, manufacturing, product development, product testing, scheduling, installation, manufacturing, and software renewal. Automation is one way to make the processes and workflow efficient, but automation can only go so far. Sometimes there is a need to redesign them to achieve the desired results. Therefore, they need to be periodically examined for improvements and updates.

A company should design a product that is easy and simple to install with minimal workflow and processes. However, this may not always be the case. If the company (producer) owns its branches with high involvement in the distribution network (as shown in Chapter 7), the company is responsible for installing the product. It means that optimizing the installation processes and workflow benefits the company directly.

Suppose that a company provides BAS retrofit solutions to its existing customers. To provide the best solutions, an employee from the company has to make multiple visits to understand the building configuration first and then recommend the potential options to the customer. Instead, the customer can upload the configuration file on company's secure portal that verifies the site/building details and ask a few questions for additional missing information. Using the tool to automatically obtain the data, the employee can spend a few days in advance in analyzing the situation. This way the employee can provide best options for the customer in the first (and hopefully the only) visit while saving time for the customer and the company. Similarly, instead of delivering physical media (e.g., CDs) to renew licenses and upgrade software, secure authentication techniques can be used to renew the software licenses.

9.3.2.4 Technology Adoption

A company can benefit quite a bit from the adoption and implementation of new, mature technologies into its ecosystem. Think of the new technologies as an opportunity to improve processes, productivity, efficiency, and profit margin on the products. Although there are some initial challenges to get out of the comfort zone and embrace new technologies, this may be the opportunity to obtain competitive advantage while keeping the employees and the stakeholders motivated. In general, there are many possible technologies, which has potential to affect many areas of a business in short-term and long-term periods; the following list provides these advanced technologies.

- Big Data
- Cloud
- Machine Learning
- AI
- Blockchain
- Distributed Computing
- IoT
- Advanced Analytics
- Smart Devices
- Virtualization
- Low-power Printable Sensors
- Embedded Systems
- Control and Robotics
- Autonomous Vehicles
- Additive Manufacturing
- Augmented Reality
- Cybersecurity
- Social Media
- 3-D Printing
- Advanced Materials
- Resources Optimization
- Hyperloop
- Service Architecture
- Industry 4.0
- Long-distance Wireless and Networking

Only some of the technologies are applicable in the area of building controls. Selection and adoption of the technologies depend on the product type, business workflow, company's culture and vision. For example, the Internet and social media technologies combined with AI are inexpensive ways to market the product to a certain group of people such as technology-savvy professionals and millennials. Another use case is on continuous delivery of selected control software with minimal human intervention. If a new version of control algorithm is developed and launched, the applications running on the cloud in real-time can determine the systems where the control algorithms are applicable. The algorithms can be deployed remotely with no or minimal human intervention using the cloud technologies. This method reduces the installation cost and upgradation cost with minimal disruptions in the working control system. Furthermore, the process is simple and convenient from the customers point of view. Another use case is on information dissemination and providing training support on new products and upgrades using smart devices, virtualization, cloud, and augmented reality. There are enormous opportunities using the technologies. Therefore, the recommendation here is to consider such technologies as substitutes to accelerate toward the company's ambitions and vision.

9.4   Supply Chain Management

Supply chain management (SCM) is essential for the success of a business because SCM acts as a lever to control the price and delivery of a product, both of which are vital attributes from customers perspectives, especially in consumable markets. There are two main types of SCM Strategies:

1. Speed-oriented: In this strategy, the main focus is on delivering the product as fast as possible to the customer reducing the total lead time, i.e., the time needed from manufacturing a product to delivering the product. This strategy requires better prediction of demand and higher inventory stocks at all times. Therefore, the strategy is robust to the uncertainty and fluctuations in the demand.

2. Price-oriented: Price-oriented SCM strategy's goal is to reduce the cost of product as much as possible so that they can stay competitive in the low-price product market. The market in this case is usually the masses as the profit margins are low because of low prices. The inventory level is optimized to reduce the product cost and product lead time may suffer in case of high oscillations in the demand.

Again, selection of SCM strategy is based on the product and marketing strategies. It is important to ensure that the SCM strategy is well aligned with the corporate and business strategies. This section discusses the effect of product/marketing strategies on supply chain decisions, and the implications of those decisions. SCM is mostly applicable to hardware products. Figure 9.5 shows a typical workflow in SCM, which consists of core components and enabling components. The workflow can vary slightly based on the type of product and business. Core components are needed for any supply chain while enabling components make the supply chain efficient and effective. The components are explained next.

- Planning: It is an activity to forecast supply and demand along with a strategy and a timeline to meet the demand, which may include other necessary arrangements. It is usually the first activity in the supply chain management.

- Procurement: Procurement is an activity to purchase raw material or an assembled product/component from a supplier.

- Production: It is the process of manufacturing products using raw materials or assembling the third-party components together. If no further modifications are needed, the products are packaged and labeled at the facility.

- Transportation: The next step is to move the final product from production facility to a warehouse. Depending on the location of the facility and supply chain strategy, the company chooses the transport mechanisms as flights, railways, roads, ships, or a combination of them.

- Warehousing and Inventory Management: Transported products are stored in warehouses. If the company is large, this could be a central warehouse of the company where the inventory of products is managed and optimized.

- Distribution: Products from the central warehouse is distributed to individual warehouses dispersed in different parts of the country or cities including small storage location in stores. The goal of a price-focused supply chain strategy in a distribution center is to ship the items as soon as they arrive the warehouse without stocking them. This requires predicting the requirements better from its customers and understanding the constraints and limitations from the supplying side, especially when dealing with high-volume customers.

- Delivery: Products are delivered from the distribution centers or warehouses to the customers. Instead of just warehouses that are used as storage facilities, distribution centers are used if some final touches or custom modifications are needed at the end before the product is delivered to the customer.

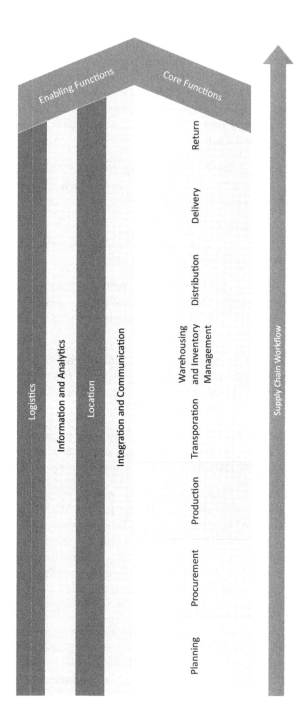

FIGURE 9.5
Supply chain management workflow with its core and enabling functions.

- Return: If a product needs to be returned for any reason, the same process may be followed backward either to store the product at a local storage facility or to send back to the central location. If the product turns out to be defective or malfunctioning, it is returned back to the production facility or the development team for further analysis so that the defect rate can be reduced in the future.

- Logistics: Logistics helps in deciding the physical flow of the material (raw material, product) to ensure responsiveness while reducing cost at the same time. It also determines the leaving time from the production facility, holding time at the storage or distribution facilities, and the delivery time of product to customers or intermediate site.

- Location: The most-effective locations of facilities for production, storage, and distribution center are identified as part of this function. Supply chain strategy and long-term corporate strategy play a key role in such decisions as it is difficult to change the locations of these facilities. For example, Walmart opened its large stores close to suburban, low-rent areas. The stores are close to their distribution centers, which are close to airports for easy and fast delivery. The original strategy of Walmart was to purchase the items in bulk and sell them at consistent low prices while reducing the total cost associated with the SCM and operations [12].

- Information and Analytics: Information is necessary to strengthen the supply chain. The question becomes: what data to collect and how to combine the data together to generate useful and meaningful information? One way to realize this vision is by having stronger IT infrastructure development that allows better visibility into multiple systems across the board. Integration of software tools and analytics platforms with internal investment will facilitate the information that aids business decisions while reducing wastage. As an example, Walmart and Apple invest heavily on information systems and technology for several reasons. Apple uses electronic devices to maintain secrecy throughout the assembly process, whereas Walmart uses RFID tags to track the inventory and merchandise location (storage vs. on the shelf) [12].

- Integration and Communication: Company can act a trusted partner by satisfying the need of customers in a timely fashion by automating the vendor management system [12]. It requires a better forecasting system on the customer requirements along with the integration of customer inventory system into the company's IT infrastructure. This way the company can plan better for the future and also provide products to the customers as fast as possible. If implemented right, the company can even supply items to its customers at lower cost by incorporating the fluctuations in the raw material price.

Some businesses models are purely based on effective supply-chains. Because of the construction bidding process and life-cycle of buildings, equipment for new buildings are ordered during the proposal phase, which happens several weeks (or months) before the products are physically installed. There is plenty of time available for the manufacturer or contractors to have the products delivered to the site. Therefore, the cost and technology-features are highly important, and thus effective supply-chain alone is not sufficient for the control manufacturers. However, better SCM can certainly provide the companies a huge competitive advantage.

In case of HVAC industry, the supply chain management involving—equipment, devices, controller, and accessories—is complex with many manual steps and processes because of the involvement of several stakeholders. The products are usually manufactured overseas. They are shipped into the US (United States) to the respective warehouses of companies or to the customers in some cases. From the stocked inventory at the warehouses, the products are directly shipped to the customers. The origin, manufacturing location, and warehouse of the product vary because many companies use a variety of products, which also include rebranded products from other global companies. In the commercial HVAC industry, which is mostly B2B for the company, the primary customers are the businesses who sell the products further to other businesses or end-customers. For their own products (i.e., non-rebranded products), companies typically have high control on the supply chain process from procuring the parts to manufacturing to transportation into the country. However, for the rebranded products, the companies have very limited control on the supply chain. In many cases, the rebranded products are directly shipped to the B2B partners to reduce the cost. In other words, the rebranded products may not be stocked.

9.4.1 Supply Chain Management Example: Apple

Apple is considered to have one of the best and most effective supply chains in the world. This example shows how a well-managed supply chain can serve as a competitive advantage while strengthening the corporate strategies. The original inventory system of Apple (30–40 years back) was not well organized, causing excessive and no inventory on multiple occasions. Apple now has more control over its supply chain than its competitors. Apple provides upfront capital investments in advance to its suppliers to secure the items for low prices in large volumes. This puts Apple in a high priority list (reducing the uncertainty and providing robustness during busy/high-demand seasons) as compared to its competitors as they don't provide upfront investments. Apple had a program for its suppliers to purchase capital equipment in exchange for supply and cost targets' assurance, e.g., agreement with a U.S. laser equipment supplier to secure hundreds of machines. Apple has much better inventory turns and sells 70% of its products directly to consumers or businesses, which again provide better control and management over its supply chain [11]. A few key attributes associated with the supply chain operations of Apple are:

- Internal management and coordination: Apple's R&D team in the US works closely with the production team overseas to ensure that the new designs can be produced in large quantity over a short period of time. Apple does most of the supply chain internally including the management as compared to other competitors. Apple designers work with suppliers and manufacturers to build strong relationships while communicating them the requirements. This way Apple has better control over the supply chain.

- Limited products: Only limited configurations and product options are available to smoothen the supply chain process. However, Apple started to provide more custom options in the past few years.

- Demand forecasting system: Apple forecasts its demand 150 days in advance and updates its forecast continuously based on sales targets or other relevant updates. These notifications help the factory to adjust their production rate accordingly. Apple monitors its store sales every hour and updates its production forecasts by the day.

- High flexibility with global suppliers: Apple enjoys high flexibility with global supplier because of its unique and exclusive negotiation terms. If many engineers and other workers in a short period of time during its supply chain functions overseas, they could be arranged easily by a global supplier. Apple uses technology (e.g., electronic devices) to maintain secrecy throughout the assembly process.

- Quick response to the demands: Apple supply chain strategy is responsive. As mentioned earlier, Apple offers upfront payments and programs to purchase equipment for the suppliers. Pre-purchasing of holiday air-freight was another step toward increasing responsiveness for Apple products. Apple provides a better customer experience by improving the responsive time and providing rapid service through reverse logistics function.

- Low Inventory Costs: Apple reduces the inventory cost by shipping the products directly to customers from the assembly plants in China.

While Apple enjoys the competitive advantage in its supply chain, it faces or may face new challenges in the future:

1. Maintain Responsiveness and Competition: Staying on top of competition to maintain responsiveness and best supply chain processes is one of the biggest challenges for the company. For example, Amazon, Dell, or any other competitor can also offer the same suppliers or transporters a higher margin deal with advanced upfront payments. Apple has to continue to invest further into the processes and strengthen the relationships with its partners.

2. Protect Secrets: Apple has/had over 150 suppliers. It is very hard to maintain secrecy throughout this global process. Although Apple uses technology (e.g., electronic devices) to maintain secrecy throughout the assembly process, it might have to update the processes or reduce the number of suppliers or reduce the number of supplier changes in the future.

3. Limited Suppliers: A few suppliers have rejected an offer from Apple to maintain their flexibility and freedom [11]. Because of diverse and specialized needs of Apple, it can be challenging to find a replacement supplier in certain conditions. Apple needs to have a backup plan and work collaboratively with such suppliers. One such weak link in the supply chain can disrupt the entire process.

In general, Apple has well-coordinated and connected supply chain processes with strong relationships established with hundreds of suppliers including other operational vendors (e.g., assemblers and transporters) across the globe. Apple supply chain strategy is unique and efficient offering several advantages over its competitors. In the future, Apple can leverage some of its existing suppliers and existing supply chain processes/infrastructure for its upcoming products. This will yield high response time and high profit margin on the products. With increasing financial numbers, new products in the pipeline, customer-focused products, responsive supply chain strategy, and an efficient supply chain process, Apple shows high growth prospects. It is recommended that Apple should continue to invest into new products to satisfy customer needs and improve its supply chain because of high competition. Apple also should work on reducing the number of dependencies on external suppliers. Recently, Apple initiated a major action of producing its own chipset instead of using Intel, which had been a supplier of Apple for decades. This was done for many reasons, and better control over supply chain is one of them.

9.4.2   Supply Chain Evolution: 1.0 to 5.0

With the advent of new technologies, the supply chain had been progressing toward 4.0, in general. Several articles have shown the progression of industry from 1.0 to 4.0 bringing in multiple perspectives [1, 5, 6]. Industry 4.0 corresponds to the use of latest technologies in the manufacturing and supply chain world. These technologies are going to dominate the industry further in upcoming years. Some of these technologies are also leveraged in an initiative by Japan, Society 5.0 [15], which is aimed to improve the human life (or the overall society) through these technologies. A glimpse of Industry 5.0 is also provided in these articles [5, 16, 17]. Figure 9.6 shows the evolution of industry from Industry 1.0 to Industry 5.0.

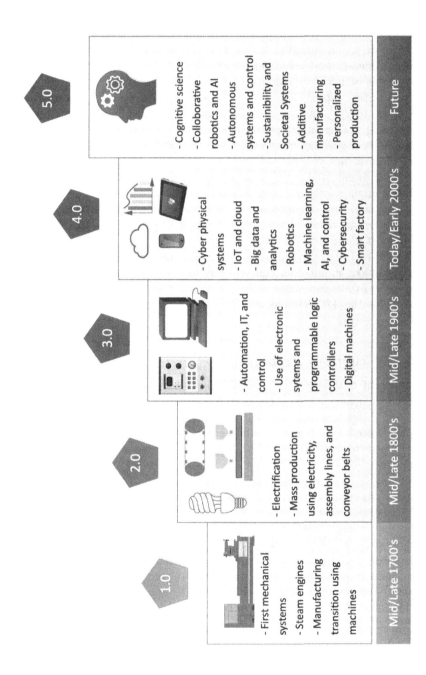

FIGURE 9.6
Industry evolution from Industry 1.0 to Industry 5.0.

9.4.3   Supply Chain Improvement Opportunities

In the building industry, there are several opportunities for the 4.0 technologies to enter in this supply chain and improve the processes. Horizontal integration [20] is an important step in providing visibility to multiple stakeholders (brokers, distributors, and mechanical contractors) during the entire process. Since the equipment delivery time is quite long in the order of several weeks to months, better forecasting and planning will improve the responsiveness and customer satisfaction. Although it seems that there is less scope of additive manufacturing, additive manufacturing combined with 3D-printing can be explored for certain components, e.g., product casing or faster prototype development. Big data, IoT, and analytics can be used to determine the right supplier, real-time location of supplier parts, and current production stage of an equipment. Because the existing supply chain management in not the state-of-art for most companies, it is reasonable to assume that augmented reality and robotics will not be the first ones to enter the supply chain management in this area.

On contrast, some form of automation will gain sufficient traction to reap the most obvious benefits, e.g., automatic payments and orders processing. The foremost and most important investment needed, for the 4.0 technologies to become a reality in this industry, is the software/infrastructure development for data storage, monitoring, and visualization at one central place, possibly in the cloud in a secure fashion. In a nutshell, there are sufficient promising occasions for the 4.0 technologies to secure an entree in the existing supply chain management.

9.4.4   Disruptions Preparation

It is apparent that several 4.0 technologies are going to disrupt the existing SCM and their associated processes in the buildings industry in the upcoming few years. A few technologies will enter faster than the other technologies. Companies need to develop transition strategies (along with the priorities) so that the new technologies can be easily incorporated into the supply chain management.

Companies should explore and invest resources to create the transition strategies. As part of the transition, an immediate step is to create a cloud-based infrastructure to gather data on supply chain activities (e.g., supplier, inventory, transportation, production, planning, forecast) so that the information is accessible worldwide. For global companies, this information can be used for several improvements and analytics. Companies have to ensure that they have acquired the right talent (control engineers, data scientists, software engineers, cyber-security experts, and software architects) and cloud storage space (in-house or well-established cloud vendors) ahead of the implementation.

A low hanging fruit will be to introduce automation in processing paperwork and payments. It means that a few processes will become obsolete with the adoption of new technologies. Companies need to recognize these processes in advance. Moreover, the new roles and responsibilities of the staff members (or entire teams) can be refined accordingly. In some cases, introduction of new technologies can result into restructuring of groups or teams. A certain set of training and professional development courses may be needed so that the staff members can easily handle new roles and responsibilities. Since a few process and experts in those roles are no longer needed, knowledge transfer may be required for the automation processes and the new staff members. Seminars and town-hall meetings are very useful to convey the message and help employees understand the reasons behind the changes. The company needs to be well-prepared so that the transition is as smooth as possible.

Because there are so many opportunities to explore and implement from the SCM perspective, it is important to evaluate the impact of these technologies on the company. A small R&D (research and development) project or team could be highly beneficial to calculate the ROI for different technologies. Environmental trends and micro-factors need to be considered during the analysis. This way the company can prioritize the investments as part of the transitioning strategy. It is also possible that some technologies are not feasible for the company today, but they could fit the business model as the company starts incorporating new technologies. Therefore, the company needs to continuously keep an eye on the growing trends, technologies, and startups. SWOT analysis is another way to evaluate the company's strategy and its direction on industry 4.0, as mentioned in Chapter 8. In short, to prepare for the industry 4.0 disruption, the company should realize the importance of industry 4.0 and the company must have a dedicated investment (via a specialized project or a separate team) to explore the opportunities and adopt the technologies that benefit the company while supporting its other strategies such as marketing, corporate, IT, and supply chain. This way the company can lower its operating costs and flatten the negative effect of these disruptions.

9.4.5 Industrial Improvement and Impact

Moving from Industry 3.0 to Industry 4.0 will provide a new look to the company that is aggressive and ambitious to move in that direction. Considering the adoption of almost all the 4.0 technologies (big data and analytics, cloud computing, system integration, cyber-security, IoT, system integration, additive manufacturing, and augmented reality) into the company operations, there will be several changes and improvements in the company in addition to the initial capital investment:

- Use of cloud computing and prediction models that use historical data and current orders to forecast the future demand and place an equipment order automatically to meet the demand at the forecasted time. At the same time, the system calculates the shipping location, route, and stocking location if needed in optimal fashion. The company will use an optimized system to stock the items in multiple warehouses based on the predicted customer locations. The system can also consider the fluctuations in the price of materials and decide the optimal time to procure or produce the equipment. For example, the demand of electronics a few months before the holidays is much higher than the demand during shoulder months.

- New inexpensive sensors are installed at multiple stages of the supply chain. These sensors can be integrated into the cloud during horizontal integration to track and increase visibility of building equipment, controllers, and accessories. As many companies sell the rebranded products of other companies, the companies will have access to the vendor inventory management system with real-time tracking, traceability, and updates. The vendor management system will be included into the company horizontal integration platform.

- A very high percentage of returned equipment are not defective in building controls. Accessibility and delivery of testing data and reports in digital form will improve both time and resources for a company and its customers in diagnosing the problems.

- Robotics will be used to move, stock, organize, package, and retrieve some products (possibly smaller products that does not require significant customization).

- Sample products will be created with digital twins. People in the US will be able to create a sample directly in the factories overseas. Someone in the factories can collect the sample and ship it to the office in the US. This will reduce both time and manpower significantly.

Industrial transformation will lead to improvements in the following operational metrics:

1. Increase productivity at the manufacturing and inventory management system.

2. Horizontal integration will increase the total customer satisfaction because of their visibility into the entire process.

3. Reduced manpower as many obsolete processes will be eliminated.

4. Since most of the processes will be automated, the asset utilization will increase.

5. Lower operating cost and reduction of operational inefficiencies. This is a huge competitive advantage similar to Japanese automotive industry in the late 60's [18].

6. Higher profit margin because of low operating costs.

7. Lowering the product price (even with same profit margin) can provide competitive advantage to yield higher sales and potentially open new doors of opportunities.

Supply chain management is a very important part of any business as it affects the operating cost and quality (responsiveness) of the products, which have direct impact on the profitability, productivity, and customer experience. Supply chain management strategy is as critical as product marketing, management, IT, and financial strategies. To have a holistic picture of a business, it is essential to look at the different functions and ensure that the processes in the SCM are optimized as well. If supply chain is ignored, it could be the weakest link in the chain. For some businesses, supply chain can be one of the most critical functions where the entire business is heavily dependent on the effectiveness of their SCM, e.g., Chipotle, while other businesses can use effective supply chain as a competitive advantage. Risks and other factors (responsiveness, etc.) need to be examined into the design of supply chain strategies.

"Beware of little expenses. A small leak will sink a great ship."

Franklin, *Benjamin*

Key Takeaways: A Few Points to Remember

1. Most firms exist to increase their profitability.

2. Financial and cost management need to be taken heavily into problem formulation for both short-term and long-term success of a company.

3. Rate of return, interest rate, present value, future value, and time value of money are building blocks upon which financial concepts and decisions are made.

4. SPP and ROI are simple and easy (and thus commonly used) methods in the building industry to make quick investment decisions while selecting a product, a project, or a portfolio of projects.

5. Net present value—an advanced, more accurate than the simple payback period—evaluates project investments while incorporating the time value of money concepts.

6. Effective cost and financial management, besides increasing the revenue, is vital to improve the firm profitability.

7. Reduction in the COGS sold (primarily for hardware products) and operating cost (primarily for software/control algorithm development) are two main ways to lower the overall cost of the product while maintaining its quality.

8. Strong SCM is a key to not only reduce the cost of hardware controller but also ensure the speed, reliability, and quality of products when procuring them from the suppliers.

9. Digital transformation and adoption of industry 4.0/5.0 technologies can help companies gain competitive advantage and reduce a variety of costs.

10. Automation, outsourcing, workflow optimization, and use of new technologies can result in efficient internal and external processes. However, their trade-offs should be carefully examined.

Bibliography

[1] The Industrial Revolution: From Industry 1.0 to Industry 4.0. https://www.seekmomentum.com/blog/manufacturing/the-evolution-of-industry-from-1-to-4, Oct 2019. Accessed: 2020-04-22.

[2] Current US inflation rates: 2009–2020. https://www.usinflationcalculator.com/inflation/current-inflation-rates/, Mar 2020. Accessed: 2020-03-22.

[3] Joseph Berk. *Cost reduction and optimization for manufacturing and industrial companies*, volume 2. John Wiley & Sons, 2010.

[4] S.M. Bragg. *Cost Reduction Analysis: Tools and Strategies*. Wiley Corporate F&A. Wiley, 2010.

[5] Kadir Demir, Gozde Doven, and Bulent Sezen. Industry 5.0 and human-robot co-working. *Procedia Computer Science*, 158:688–695, 01 2019.

[6] Joaquin Fuentes-Pila, Jose Garcia, Arianna Latini, Carlo Campiotti, Dina Murcho, F. Baptista, Luis Silva, Jose Marques Da Silva, Jose Marques,

and Germina Giagnacovo. D.6.7. Best practices for improving energy efficiency. Technical report, Tesla Project, 11 2015.

[7] Amy Gallo. A refresher on internal rate of return. *Harvard Business Review*, 03 2016.

[8] Gridconnect. ESP8266EX - Tiny Wireless 802.11 B/G/N Chip. `https://www.gridconnect.com/products/esp8266ex-tiny-wireless-802-11-b-g-n-chip`. Accessed: 2020-04-02.

[9] Adam Hayes. Benefit-Cost Ratio (BCR). `https://www.investopedia.com/terms/b/bcr.asp/`, 11 2019. Accessed: 2020-03-22.

[10] Asio American Inc. 1000ft CAT5E plenum cmp rated cable - white. `https://www.bncables.com/cat5e-bulk-cables/cat5e-cmp-plenum-utp/bulk-cat5e-plenum-cmp-rated-cable-white.html`. Accessed: 2020-04-02.

[11] Fraser P. Johnson and Ken Mark. Apple Inc.: Managing a Global Supply Chain. *Ivy Publishing*, pages 1–21, 03 2014.

[12] P. Fraser Johnson and Ken Mark. Half a Century of Supply Chain Management at Wal-Mart. *Harvard Business Review*, 2012.

[13] Steven A. Y. Lin. The modified internal rate of return and investment criterion. *The Engineering Economist*, 21(4):237–247, 1976.

[14] B.E. Needles, M. Powers, and S.V. Crosson. *Principles of Accounting*. Cengage Learning, 2010.

[15] Government of Japan. Society 5.0. `https://www8.cao.go.jp/cstp/english/society5_0/index.html`. Accessed: 2020-03-22.

[16] Esben H. Ostergaard. Welcome to industry 5.0. `https://ww2.isa.org/intech/20180403/`. Accessed: 2020-04-02.

[17] Banu Ozkeser. Lean innovation approach in industry 5.0. In *International Conference on Research in Education*, 04 2018.

[18] Michael E. Porter. What Is Strategy? *Harvard Business Review*, 1996.

[19] William Stevenson. *Operation Management*. McGraw-Hill, USA, 2009.

[20] Sohrab Vossoughi. Today's Best Companies are Horizontally Integrated. *Harvard Business Review*, 2012.

10

Controls Business Framework

Previous chapters have been packed covering many technical and business topics related to the building controls industry. Although these topics are very informative, they omit certain details because of their diversity and the scope of the book. This chapter aims to provide a holistic picture to the readers and help them connect the dots through the knowledge they have gained from this book. This way, one can understand his/her role in entire workflow and contribute positively.

10.1 Combined Technical and Business Perspective

Controls and applications in buildings offer endless opportunities from both business and technical perspectives, especially when both perspectives are merged together to complement each other. The following are the main reasons to justify the endless opportunities:

- As discussed in the previous chapters, the state-of-art technologies used in most building systems/components are way behind the technologies in most progressive industries. It means that the building companies have the opportunities to explore the technologies and apply them in the building sector, e.g., cloud integration into BAS. Since the technologies have been validated in other sectors, the heavy lifting is already done and the building control companies can leverage them for their business cases and applications.

- To gain competitive advantage in the building controls area through technical superiority, the companies have to invest into research programs or possibly collaborate with research institutions. There are tremendous research opportunities in this arena although the initial investment could be very high. For example, there are almost three dozen of control algorithms, but only a handful of them either exist in buildings or are being extensively researched. In short, there are many unexplored technical areas both at the component level and the system level.

- Customer needs and requirements have been changing at much faster rate than before. Millennials have different values and preferences as compared to baby boomers. Furthermore, the volatility in environment, globalization, government policies, regulations, and company structures add another layer of complexity. It means that the traditional approaches to solve customer problems may not necessarily apply. New business models and revenue streams are expected to emerge. The most resilient companies [1] will thrive despite the entry of new market players.

It is not that people in the industry are not aware of these opportunities. In fact, building industry had realized it a long time ago and thus many large companies have tried (or explored) different products and business models to tackle the challenges and convert them into opportunities. However, none of them have been able to capitalize on the opportunity and claim the victory in a significant manner. Figure 10.1 shows a proposed framework for control business in intelligent buildings. The figure shows an end-to-end workflow of high-level processes that have been covered throughout the previous chapters.

10.1.1 Strategy and Marketing Management

The first process is the strategy and marketing management, which had been covered in Chapters 7 and 8. Basically, the outcome of this step is detailed business and product strategy for the company. Supporting the mission and vision of the company, strategy is a set of activities/attributes to create a unique system that provides long-term, superior profitability to the company. As shown in the article [1], strategy is not just implementing a random list of specific activities or all possible plans. Instead, it is about choosing the right set of activities (and ignoring the rest of them) that align well with the company's (or a division's) mission, vision, values, and objectives. Product and marketing management goes deeper into identifying and satisfying the needs of customers through specific products, services, experiences, and information. It includes creating, communicating, and delivering value to the customers. Strategy is important for several reasons as it [2]:

1. Creates a unique value while providing significant competitive advantage to a company to drive its profitability

2. Places the company in a position that competitors are not able to replicate for a long period of time

3. Defines/limits the scope and boundaries on business the activities, e.g., the targeted market, customer type in the targeted market, and services/products for the targeted market

4. Provides guidance on individual activities (functional, product-level, marketing, etc.) and tactics

The outcome of this step is a set of strategies depicting the company and

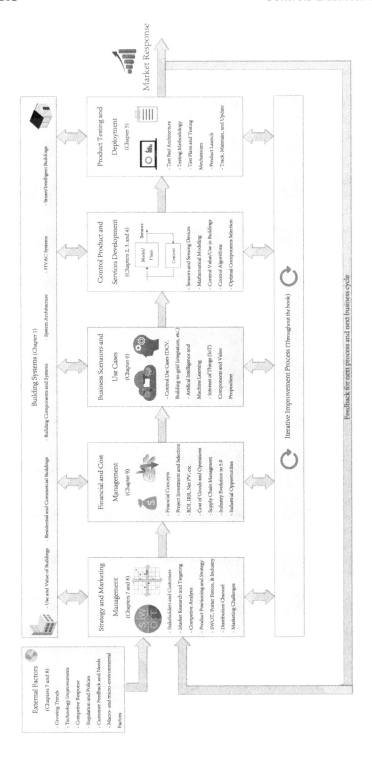

FIGURE 10.1

Framework for controls business in intelligent buildings.

division direction at high level. Sometimes, this is referred to as business-level requirements for a certain product/business positioning. SWOT, competitive analysis and market challenges are a few activities performed during the analysis.

10.1.2 Financial and Cost Management

The business-level requirements can be satisfied through different projects or multiple options within a single project. Financial and cost management concepts, described in Chapter 9, are used to make the decisions that will produce a better outcome financially. A few topics in this category are ROI, IRR, SPP, rate of return, inflation rate, PV, FV, NPV, and time value of money. Other concepts on cost management are reduction in COGS and operational expenses. For example, suppose it is finalized that a hardware controller with certain features and attributes need to be manufactured. Now, the same parts of the controller are sold by many manufacturers. Each manufacturer has a unique value proposition with different terms and conditions. The question arises: who should be chosen as the supplier? The concepts discussed in the chapter help in answering such questions from financial/cost point-of-view. SCM and advanced industrial technologies are potential contributors to overall cost reduction while increasing the responsiveness. The steps here ensure efficient operation of the plan. Other non-technical aspects of the plan are validated and verified in greater detail such as finances, project selection, and procurement.

10.1.3 Business Scenarios and Use Cases

This step, explained in Chapter 6, bridges the gap between technical and business needs. The products and use cases are defined at higher level with sufficient details that the customers can understand and relate to. A group of these use cases can constitute the feature list of a product or products. For example, DCV is a technical term that the customers can comprehend immediately but not the business or financial leaders although there are business benefits in terms of energy savings and better IAQ. People involved in this step are technical leads and product/portfolio managers.

Selection of technologies can also be included in this process. AI and IoT are optional technologies which may not be needed to accomplish the features. However, use of such technologies may provide competitive advantage in other areas such as marketing and technical excellence. Many use cases and technologies (AI, ML, IoT) are briefly explained in Chapter 6.

10.1.4 Control Product/Services Development

Chapters 2, 3, and 4 discuss the technical details and options during controls product development. As there are many choices during the development cycle, choosing the right system is important to satisfy the business needs accurately.

Depending on the type of control product (hardware, software, service, etc.), its development can incorporate sensing, mathematical modeling, and control-focused components such as algorithms and architecture. Designing the right system means selection of several components correctly. Product stages, design processes, technology options, and tradeoffs between different technical choices are also discussed in these chapters.

10.1.5   Product Testing and Deployment

Once the product is developed, it needs to be tested before hitting the market. Depending on the testing approach, product testing can be done during different stages of product development. At the end of successful testing, the product is technically ready to be launched. If a problem is detected during the testing phase, the development teams get involved to resolve the issue. Creating a test bed, choosing a testing methodology, and development of test plans are parts of this large activity. After the product is launched, the next steps are to track, maintain, and update the product. This may include bug fixes, patches, and minor enhancements until the next major version of the product is launched into the market. Chapter 5 explains the product testing and deployment steps.

10.1.6   Building Systems and Iterative Improvement

Types of buildings and building components along with their working and control system architecture are described in Chapter 1. Chapter 1 also explores the question of what does it mean for a building to be smart and intelligent. It is clear from the figure that knowledge of building systems is used in almost every part of the framework. The major difference is that not everyone needs the same information. For example, a business leader needs to know the types of building/systems and their principal activity so that he/she can develop the positioning strategy. On the other hand, a technical lead or a controls engineer must understand the working of different building components and their architecture. Therefore, the information present in Chapter 1 is utilized throughout the book.

Another very important part of this framework is iterative improvement process, which is needed to share information and provide feedback between different activities and thought processes. With modern technologies and thinking, it is strongly recommended to have continuous interactions without spending too much time on it. It is very challenging to have continuous interactions while making them quick, productive and useful. Transparency and cross-functional knowledge of people involved during different parts of business are essential for this. Higher the knowledge and transparency level, the less is the time consumed during the iterative improvement process. Basically, this step acts as a validation mechanism to bridge the gaps between technical and business perspectives. With this approach, the entire process becomes highly agile so that we can avoid big surprises or major rerouting later down the road.

10.1.7   Market Response and External Factors

Once the product is launched into the market, the next big major step is to capture the response of the market. Market response includes feedback from different stakeholders and the product evaluation such as the value of product in the eyes of its customers, revenue/sales, and competitor response. An action is accordingly taken based on the response. Sometimes, the action could be revamping the marketing activities only with no change in the technical side of the product. Details on some marketing activities are provided in Chapter 8. In other scenarios, it may require the company to change its business model and update the product(s).

External factors are also analyzed as part of the feedback for the next business cycle because many external changes had occurred since the inception of marketing/product strategy. It is possible that new technologies may have been invented, or the competitors have launched new competing products, or the government may have changed the policies or added tariffs on exports since the original strategy development. The external factors are market trends, technology improvements, competitors' response, new regulations and government policies, changing customer needs, and other macro/micro-environment factors. These factors are discussed in Chapters 7 and 8. In the most essential respects, market response and external factors are clubbed together to refine and update the strategy for the business. The entire process repeats again. It is important to emphasize that the external factors are also evaluated during the initial development of business, marketing, and product strategies.

10.2   Next-generation Building through Advanced Controls and Applications

Figure 10.2 provides an example of next-generation intelligent buildings through advancements, developments, integration, and deployment of technologies in advanced controls and applications for building systems. There are four major areas that highlight the concept of intelligent building and provide a complete picture to the readers:

- Benefits: The benefits of intelligent buildings include energy savings, less wastage of resources such as water and gas, reduce carbon emission because of better control and less energy consumption/wastage, better indoor climate for occupants, improved efficiency of operations and equipment, lower cost (maintenance, ownership, and capital) because of predictive techniques and modeling, improved security and safety of systems because of better algorithms, and improved resiliency/robustness and accuracy of the systems as they can better handle uncertain situations, e.g., fault-tolerant control techniques implemented in case of sensor failure.

• Beneficiaries: Depending on the level and type of use cases, there are many groups benefiting from intelligent buildings. Although the primary beneficiaries are occupants and building owners/operators, other stakeholders will reap the benefits directly or indirectly at different levels. Note that it is critical to create an ecosystem and product that benefit the secondary stakeholders because without their involvement it is very challenging the make the next-generation buildings a reality.

• Use Cases: The use cases are the high-level scenarios that will achieve different benefits for various stakeholders. A few use cases are plug-n-play devices for low installation cost, FDD for low maintenance cost, autonomous buildings with minimal human involvement for low ownership cost, smart energy conservation techniques for economic savings, remote monitoring and control for easier access, smart alarming and visualizations for simple insights of the system, user behavior and intervention for energy savings and better productivity, integrated systems and unified interfaces for better productivity and operational efficiency, precise personalized control for healthy indoor climate, and tracking systems for safety and security purposes.

• Technologies: Technologies act as enablers and accelerators to achieve the use cases and benefits that were quite difficult or slow to accomplish otherwise. A few technologies relevant to the aforementioned use cases are AI, ML, IoT, cloud computing, wireless technologies, automation and Industry 4.0 technologies, pattern recognition, robotics, sensing, mathematical modeling, and advanced controls, which include new algorithms/approaches and efficient architectures.

As shown above, all the topics are linked to each other in a meaningful way as they may not be complete by themselves alone. Most of the topics in all the four areas are explained in detail as part of different chapters. However, each topic is so vast that it deserves an entire chapter or a book itself. Figure 10.2 shows a snapshot of buildings into a few years from today considering the existing trends and technologies. If a new business model or technology arises, they can be incorporated into the framework. It also does not mean that every potential benefit or technology will be realized or need to be realized. The major takeaway is that there are ample opportunities in the building controls arena. Since there are many challenges during this transformation, the final accomplishment or achievements on the way are highly rewarding as well. This purpose of this book is to make the fundamentals and tools accessible to the

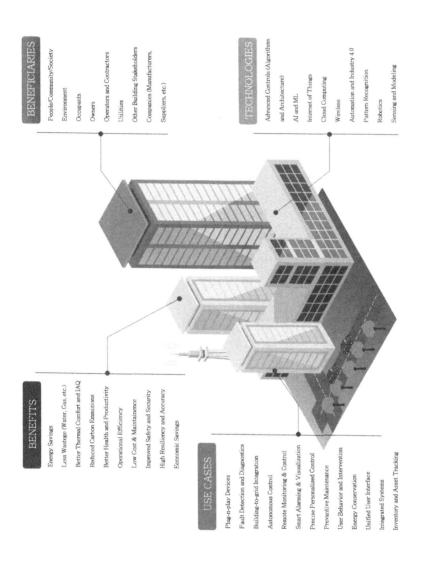

BENEFICIARIES

People/Community/Society

Environment

Occupants

Owners

Operators and Contractors

Utilities

Other Building Stakeholders

Companies (Manufacturers,

Suppliers, etc.)

TECHNOLOGIES

Advanced Controls (Algorithms

and Architecture)

AI and ML

Internet of Things

Cloud Computing

Wireless

Automation and Industry 4.0

Pattern Recognition

Robotics

Sensing and Modeling

BENEFITS

Energy Savings

Less Wastage (Water, Gas, etc.)

Better Thermal Comfort and IAQ

Reduced Carbon Emmisions

Better Health and Productivity

Operational Efficiency

Low Cost & Maintainence

Improved Safety and Security

High Resiliency and Accuracy

Economic Savings

USE CASES

Plug-n-play Devices

Fault Detection and Diagnostics

Building-to-grid Integration

Autonomous Control

Remote Monitoring & Control

Smart Alarming & Visualization

Precise Personalized Control

Prevantive Maintenance

User Behavior and Intervention

Energy Conservation

Unified User Interface

Integrated Systems

Inventory and Asset Tracking

FIGURE 10.2
Glimpse into the next-generation intelligent buildings: benefits, beneficiaries, use cases, and technologies.

readers so that they can make smart decisions in their respective fields. In the end, buildings are made for us, not the other way.

Bibliography

[1] Gary Hamel and Liisa Valikangas. The Quest for Resilience. *Harvard Business Review*, 2003.

[2] Michael E. Porter. What Is Strategy? *Harvard Business Review*, 1996.

Alphabetical Index

Printed in the United States
by Baker & Taylor Publisher Services